Analytical Profiles

of

Drug Substances

and

Excipients

EDITORIAL BOARD

Analytical Profiles

of Drug Substances

and Excipients

Volume 27

edited by

Harry G. Brittain

Center for Pharmaceutical Physics
10 Charles Road
Milford, New Jersey 08848

Founding Editor:

Klaus Florey

ACADEMIC PRESS

A Harcourt Science and Technology Company

San Diego   San Francisco   New York   Boston   London   Sydney   Tokyo

This book is printed on acid-free paper.

Academic Press
A Harcourt Science and Technology Company
525 B Street, Suite 1900, San Diego, California 92101-4495, USA
http://www.academicpress.com

Academic Press
Harcourt Place, 32 Jamestown Road, London NW1 7BY, UK
http://www.academicpress.com

International Standard Book Number: 0-12-260827-5

PRINTED IN THE UNITED STATES OF AMERICA
01   02   03   04   05   06   QW   9   8   7   6   5   4   3   2   1

CONTENTS

AFFILIATIONS OF EDITORS AND CONTRIBUTORS

Abdullah A. Al-Badr: Department of Pharmaceutical Chemistry, College of Pharmacy, King Saud University, P.O. Box 2457, Riyadh-11451, Saudi Arabia

Abdulrahman A. Al-Majed: Department of Pharmaceutical Chemistry, College of Pharmacy, King Saud University, P.O. Box 2457, Riyadh-11451, Saudi Arabia

Mahmoud M. Al-Omari: The Jordanian Pharmaceutical Manufacturing Co., Naor P.O. Box 94, Amman, Jordan

Mahmoud Ashour: The Jordanian Pharmaceutical Manufacturing Co., Naor P.O. Box 94, Amman, Jordan

Adnan A. Badwan: The Jordanian Pharmaceutical Manufacturing Co., Naor P.O. Box 94, Amman, Jordan

Harry G. Brittain: Center for Pharmaceutical Physics, 10 Charles Road, Milford, NJ 08848-1930, USA

Richard D. Bruce: McNeil Consumer Healthcare, 7050 Camp Hill Road, Fort Washington, PA 19034, USA

Nidal Daraghmeh: The Jordanian Pharmaceutical Manufacturing Co., Naor P.O. Box 94, Amman, Jordan

Alekha K. Dash: Department of Pharmaceutical & Administrative Sciences, School of Pharmacy and Allied Health Professions, Creighton University, Omaha, NE 68178, USA

William F. Elmquist: Department of Pharmaceutical Sciences, College of Pharmacy, University of Nebraska Medical Center, Omaha, NE 68198, USA

Hussein I. El-Subbagh: Department of Pharmaceutical Chemistry, College of Pharmacy, King Saud University, P.O. Box 2457, Riyadh-11451, Saudi Arabia

Klaus Florey*: 151 Loomis Court, Princeton, NJ 08540, USA

Antonio Cerezo Galán: Department of Pharmacy and Pharmaceutical Technology, Faculty of Pharmacy, University of Granada, 18071-Granada, Spain

Timothy P. Gilmor: McNeil Consumer Healthcare, 7050 Camp Hill Road, Fort Washington, PA 19034, USA

Jeffrey Grove: Laboratoires Merck Sharp & Dohme-Chibret, Centre de Recherche, Riom, France

John D. Higgins: McNeil Consumer Healthcare, 7050 Camp Hill Road, Fort Washington, PA 19034, USA

Gunawan Indrayanto*: Laboratory of Pharmaceutical Biotechnology, Faculty of Pharmacy, Airlangga University, Jl. Dharmawangsa dalam, Surabaya 60286, Indonesia

Dominic P. Ip*: Merck, Sharp, and Dohme, Building 78-210, West Point, PA 19486, USA

Krishan Kumar*: Merial Limited, 2100 Ronson Road, Iselin, NJ 08830, USA

José M. Ramos López: Scientific Instrumentation Center, University of Granada, 18071-Granada, Spain

Stephen A. Martellucci: McNeil Consumer Healthcare, 7050 Camp Hill Road, Fort Washington, PA 19034, USA

David J. Mazzo*: Preclinical Development, Hoechst Marion Roussel, Inc., Route 202-206, P.O. Box 6800, Bridgewater, NJ 08807, USA

Lina Nabulsi: The Jordanian Pharmaceutical Manufacturing Co., Naor P.O. Box 94, Amman, Jordan

Niran Nugara: Analytical Development, Schering-Plough Research Institute, Kenilworth, NJ 07033, USA

Marie-Paule Quint: Laboratoires Merck Sharp & Dohme-Chibret, Centre de Recherche, Riom, France

Isam Ismail Salem: Department of Pharmacy and Pharmaceutical Technology, Faculty of Pharmacy, University of Granada, 18071-Granada, Spain

Amal Shervington: Faculty of Pharmacy, University of Jordan, Amman, Jordan

Leroy Shervington*: Pharmacy Faculty, Applied Science University, Amman 11931, Jordan

Richard Sternal: Analytical Development, Schering-Plough Research Institute, Kenilworth, NJ 07033, USA

Reema Al-Tayyem: Faculty of Agriculture, University of Jordan, Amman, Jordan

Scott M. Thomas: Merck Research Laboratories, Rahway, NJ, USA

Timothy J. Wozniak*: Eli Lilly and Company, Lilly Corporate Center, MC-769, Indianapolis, IN 46285, USA

PREFACE

The comprehensive profile of drug substances and pharmaceutical excipients as to their physical and analytical characteristics continues to be an essential feature of drug development. The compilation and publication of comprehensive summaries of physical and chemical data, analytical methods, routes of compound preparation, degradation pathways, uses and applications, *etc.*, is a vital function to both academia and industry. It goes without saying that workers in the field require access to current state-of-the-art data, and the *Analytical Profiles* series has always provided information of the highest quality. For this reason, profiles of older compounds are updated whenever a sufficient body of new information becomes available.

The production of these volume continues to be a difficult and arduous mission, and obtaining profile contributions is becoming ever more difficult. One cannot deduce whether this is due to the new requirements of drug development to do more with less, the wide range of activities now required by professionals in the field, or the continuing personnel down-sizing, but the effect is the same. Some companies even take the near-sighted view that publishing a profile will somehow help their ultimate generic competitors. The latter concern is totally unfounded, since the publication of a drug substance profile actually sets the standard that the generic hopefuls would have to meet. The need for analytical profiles remains as strong as ever, even as potential authors become scarcer all the time. However, the contributors to the present volume have indeed found the resources to write their chapters, and I would like to take this opportunity to salute them for their dedication to this work.

As always, a complete list of available drug and excipient candidates is available from the editor. I continue to explore new and innovative ways to encourage potential authors, and welcome suggestions as to how to get people involved in the writing of analytical profiles. Communication from new and established authors is always welcome, and Email contact (address: hbrittain@earthlink.net) is encouraged. I will continue to look forward to working with the pharmaceutical community on the *Analytical Profiles of Drug Substances and Excipients*, and to providing these information summaries that are of such great importance to the field.

Harry G. Brittain

ARGININE

Amal Shervington[1] and Reema Al-Tayyem[2]

(1) Faculty of Pharmacy
University of Jordan
Amman , Jordan

(2) Faculty of Agriculture
University of Jordan
Amman , Jordan

ANALYTICAL PROFILES OF
DRUG SUBSTANCES AND EXCIPIENTS
VOLUME 27

1

Contents

1.   Description

1.1   Nomenclature

1.1.1   Chemical Name

2-amino-5-guanidinovaleric acid

(S)-2-amino-5-[(aminoiminomethyl)amino]pentanoic acid

1.1.2   Nonproprietary Names

Arginine

L-Arginine

L-(+)-Arginine

1.2   Formulae

1.2.1   Empirical

Arginine: $C_6H_{14}N_4O_2$

Arginine Hydrochloride: $C_6H_{15}N_4O_2Cl$

1.2.2   Structural

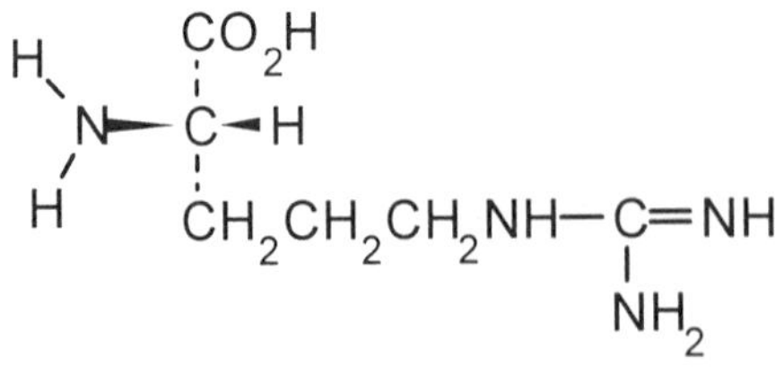

$$H\diagdown \atop H\diagup N - \underset{\underset{CH_2CH_2CH_2NH-\underset{\underset{NH_2}{|}}{C}=NH}{|}}{\overset{CO_2H}{\overset{|}{C}}}\!\!\!-H$$

1.3   Molecular Weight

Arginine: 174.202
Arginine Hydrochloride: 210.663

1.4   CAS Number

123456789

1.5   Appearance

Arginine is a white or almost white crystalline powder, obtained as practically odorless crystals.

1.6 Uses and Applications

Arginine is an amino acid that is best known as a growth hormone releaser. The decrease of growth hormone in the human body with aging is a major reason why muscle mass tends to decrease with age, and body fat tends to increase with age. Decreases in growth hormones also are partially responsible for the slower rate of skin growth with aging, which results in thinner and less flexible skin. Injections of growth hormone can reverse these problems, but there are potential dangers in receiving too much growth hormone. Growth hormone cannot be taken orally, because as a peptide, it is broken down in the digestive tract. Growth hormone injections are so expensive that few people can afford them unless they are used for a specific disease covered by insurance [1,2].

Dietary arginine supplementation (1%) of a control laboratory chow containing adequate amounts of arginine for growth and reproduction increases thymic weight, cellularity, and thymic lymphocyte blastogenesis in rats and mice [3,4]. In addition, arginine supplementation can alleviate the negative effect of trauma on these thymic parameters [5]. It has been demonstrated that arginine becomes an essential amino acid for survival and wound healing in arginine-deficient rats [6]. This work showed that 1% arginine supplementation of non-deficient rats led to decreased weight loss on the first day post-injury, and increased wound healing in rats subjected to dorsal skin wounding [6].

Arginine is also a powerful immune stimulant agent [7-9]. At one time, this was thought to be exclusively due to its growth hormone releasing properties, but arginine has been found to be a powerful immune stimulant and wound healing agent even in the absence of significant growth hormone release. Long term oral administration of L-arginine reduces intimal thickening and enhances neoendothelium-dependent acetylcholine-induced relaxation after arterial injury [10]. In addition, oral L-arginine improves interstitial cystitis symptom score [11].

Arginine is used in certain conditions accompanied by hyperammonaemia. In addition, arginine chloride has also been used as acidifying agent [12]. In severe metabolic alkalosis, intravenous doses (in gram quantities) have been calculated by multiplying the desired decrease in plasma-bicarbonate concentration (mEq per liter) by the patients body-weight (in kg) and then dividing by 9.6. In overdose, a suggested dose is 10g intravenously over 30 minutes [12].

Arginine has also been used as various salt forms, such as the acetylasparaginate, asparatate, citrate, glutamate, oxoglurate, tidiacicate, and timonacicate salts [12].

L-Arginine is a basic, genetically coded amino acid that is an essential amino acid for human development. It is a precursor of nitric oxide [13], and is synthesized by the body from ornithine. Arginine has been classified as a conditionally indispensable amino acid [14].

2.   Method of Preparation

Arginine can be synthesized from ornithine, a urea cycle intermediate [14].

3.   Physical Properties

3.1   Particle Morphology

When isolated from water, arginine is obtained as minute round crystals. A commercial sample was evaluated using optical microscopy, with the data being obtained on a Leica Diastar system.

Figure 1.   Photomicrograph of commercially obtained arginine, obtained at a magnification of 200x.

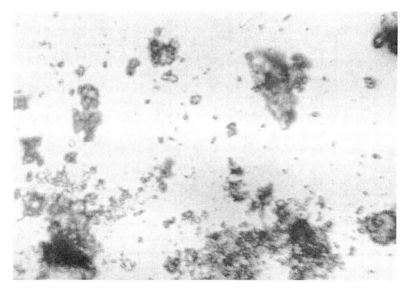

3.2   X-Ray Powder Diffraction Pattern

The x-ray powder pattern of arginine is found in Figure 2, and the table of crystallographic properties deduced from this pattern is located in Table 1.

3.3   Optical Rotation

The specific rotation of a 80 mg/mL of arginine dissolved in 6N HCl is between +26.5° and +26.9°.

3.4   Thermal Methods of analysis

3.4.1   Melting Behavior

Arginine is observed to melt at 235°C with decomposition.

3.4.2   Differential Scanning Calorimetry

The differential scanning calorimetry thermogram of arginine was obtained using DSC PL-STA Rheometric Scientific system, connected to a model No. 530000 interface. The thermogram thusly obtained is shown in Figure 3, along with the thermogravimetric analysis. The only detected thermal event was the melting endotherm at 244.62°, for which the onset temperature was found to be 243.06°. Integration of the melting endotherm yielded an enthalpy of fusion equal to 93.92 cal/g.

3.5   Hygroscopicity

Arginine is not a hygroscopic substance when exposed to ordinary environmental conditions. The compendial requirement supports this conclusion in that arginine dried at 105°C for 3 hours does not lose more than 0.5% of its weight [15].

3.6   Solubility Characteristics

Arginine is freely soluble in water (1 g dissolves in 5 mL of water), sparingly or very slightly soluble in alcohol, and practically insoluble in ether [12].

Figure 2.   X-ray powder diffraction pattern of arginine.

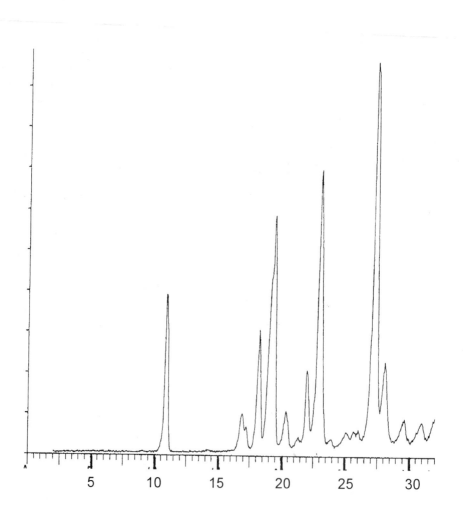

Scattering Angle (degrees 2-θ)

Table 1

Crystallographic Data from the X-Ray Powder Diffraction Pattern of Arginine

Scattering angle (degrees 2-θ)	d-spacing (Å)	Relative Intensity (%)
9.045	9.79314	0.3
10.990	8.06395	34.8
14.195	6.24965	0.5
16.930	5.24570	10.4
17.255	5.14763	6.2
18.260	4.86652	31.4
19.080	4.65918	44.8
19.405	4.58188	63.0
20.345	4.37226	11.3
21.330	4.17253	4.4
21.980	4.05059	22.7
22.980	3.87654	77.8
23.935	3.72399	3.9
25.055	3.56001	5.8
25.620	3.48277	5.5
26.060	3.42496	5.7
27.365	3.26452	100.0
28.130	3.17746	23.4
29.630	3.01994	9.5
30.995	2.88999	8.4

Figure 3.   Differential scanning calorimetry and thermogravimetric analysis thermograms of commercially obtaiend arginine.

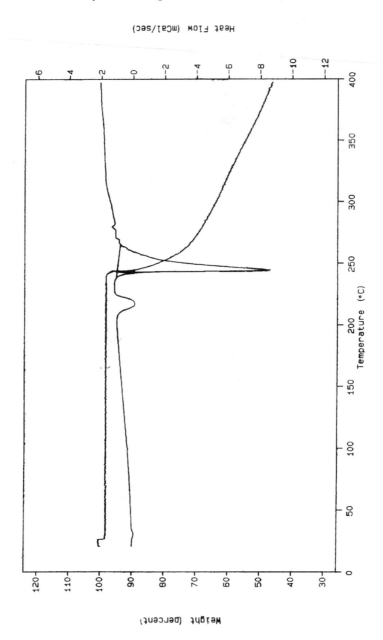

3.7   Partition Coefficient

The partition coefficient for arginine was calculated using the logP program produced by Advanced Chemical Development (Toronto, CA). The program predicted log P (octanol/water) for the neutral form to be equal to –4.08 ± 0.7, indicating a considerable degree of hydrophilicity for this compound.

The pH dependence of the calculated log D values is shown in Figure 4.

3.8   Ionization Constants

The ionization constants of arginine can be summarized as:

pK_{a1} (-COOH) = 2.17

pK_{a2} (α-NH$_3$) = 9.04

pK_a (R-group) = 12.48

The isoelectric point of arginine is found to be pH = 10.76 [14].

3.9   Spectroscopy

3.9.1   Vibrational Spectroscopy

The infrared absorption spectrum of arginine was recorded on a Jasco FTIR 300 E spectrometer, using the potassium bromide pellet method. The spectrum spanning 400 to 4000 cm^{-1} is shown in Figure 5, and assignments for the observed bands are provided in Table 2.

3.9.2   Nuclear Magnetic Resonance Spectrometry

3.9.2.1 ^{1}H-NMR Spectrum

The ^{1}H-NMR spectrum of arginine was obtained on a Bruker 300 MHz spectrometer, using deuterated water as the solvent and tetramethylsilane as the internal standard. The spectrum is shown in Figure 6, and a summary of the assignments for the observed resonance bands is provided in Table 3.

It should be noted that the protons linked to the nitrogen groups of arginine are not observed in the spectra, since they are replaced by deuterium derived from the deuterated water used as the solubilizing solvent.

Figure 4.   pH dependence of log D values calculated for arginine.

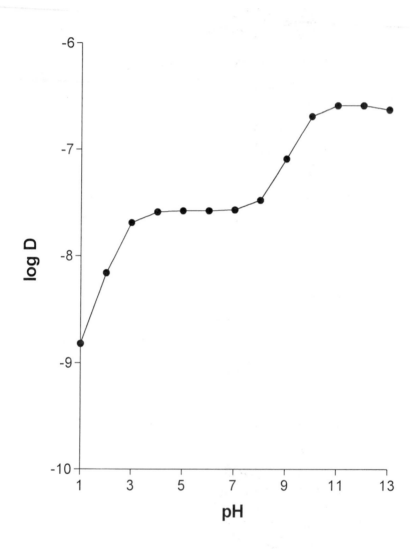

Figure 5. Infrared absorption spectrum of commercially obtained arginine, showing the bands in transmission mode.

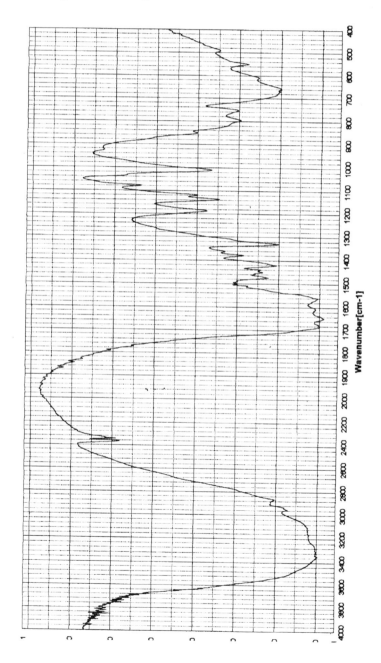

Table 2

Assignment for the Vibrational Transitions of Arginine

Energy (cm^{-1})	Assignment
3800-2400	O-H stretching mode associated with the hydroxyl groups (intramolecular hydrogen bonding of the carboxylic group)
3500-3300	N-H stretching mode of the amino group and the imine group, overlapped by the strong absorption of the carboxyl O-H group
1710-1690	C=O stretching mode of the carbonyl group
1480	C-H bending mode of the methylene groups
1150-1000	C-O stretch of the carboxylic group

Figure 6.   ¹H-NMR spectrum of commercially obtained arginine.

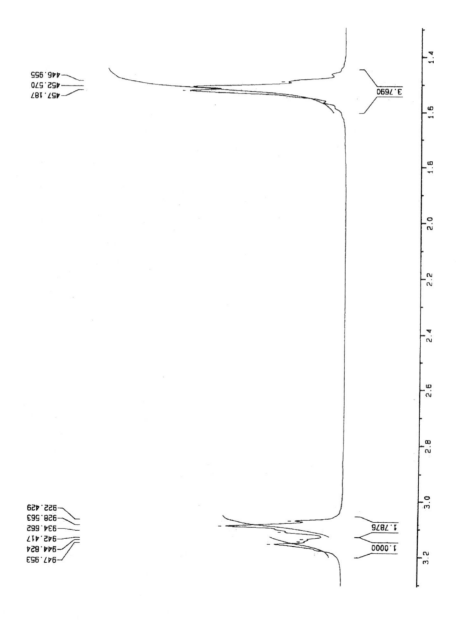

Table 3

Assignment for the Observed ^{1}H-NMR Bands of Arginine

Chemical Shift (ppm)	Number of Protons	Assignment
4.75	—	D_2O (H_2O)
3.15	1	Methine proton of the amino acid, Linked to the chiral center at -C$\underline{H}$–COOH
3.09	2	Methylene protons of the HN–C$\underline{H}_2$ group
1.52- 1.50	4	Protons of $C\underline{H}_2$–$C\underline{H}_2$–CH–COOH $\mid$ NH$_2$

3.9.2.2 ^{13}C-NMR Spectrum

The ^{13}C-NMR spectrum of arginine was also obtained in deuterated water at ambient temperature, using tetramethylsilane as the internal standard. The one-dimensional spectrum is shown in Figure 7, while the Dept-135 spectrum is shown in Figure 8. Both spectra were used to develop the correlation between chemical shifts and spectral assignment that are given in Table 4.

3.10   Micromeritic Properties

3.10.1   Bulk and Tapped Densities

The bulk density of commercially available arginine was determined by measuring the volume of known mass of powder that had been passed through a screen into a volume-measuring device, and calculating the bulk density by dividing the mass by the volume. The average bulk density of the arginine sample studied was found to be 0.572 g/mL.

The tapped density was obtained by mechanically tapping a measuring cylinder containing a known amount of sample using a Pharma Test (PT-T.D) instrument. After observing the initial volume, the cylinder was mechanically tapped, 100 times over a period of one minute. The tapped density is calculated as the mass divided by the final tapped volume, it was found that the average tapped density of the arginine sample was 0.715 g/mL.

3.10.2   Powder Flowability

The Carr Compressibility Index:

$$CI \quad = \quad 100 \; (V_o - V_f) \; / \; V_o$$

and the Hauser Ratio:

$$HR \quad = \quad V_o \; / \; V_f$$

are two values that can be used to predict the propensity of a given powder sample to be compressed. The values for V_o (original bulk volume of powder) and V_f (final tapped volume of powder) are obtained during performance of the determination of bulk and tapped density. The Compressibility Index for arginine was found to be approximately 20, indicating that this powdered sample would be predicted to exhibit fair flowability. The Hauser Ratio was determined to be 1.25, which also indicate that the powder would exhibit fair degrees of powder flow.

Figure 5.   One-dimensional ^{13}C-NMR spectrum of commercially
             obtained arginine.

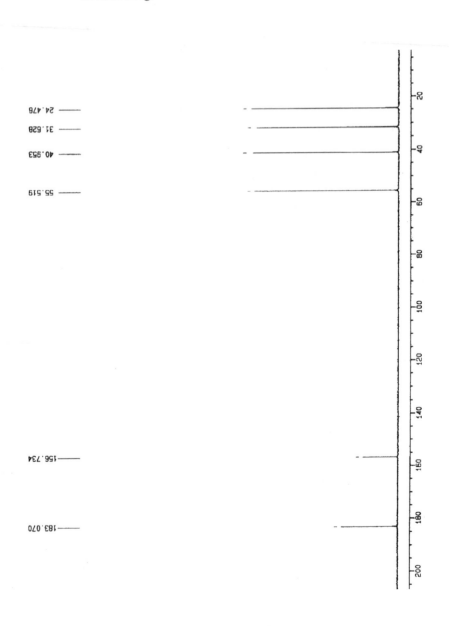

Figure 6. Dept 135 ^{13}C-NMR spectrum of commercially obtained arginine.

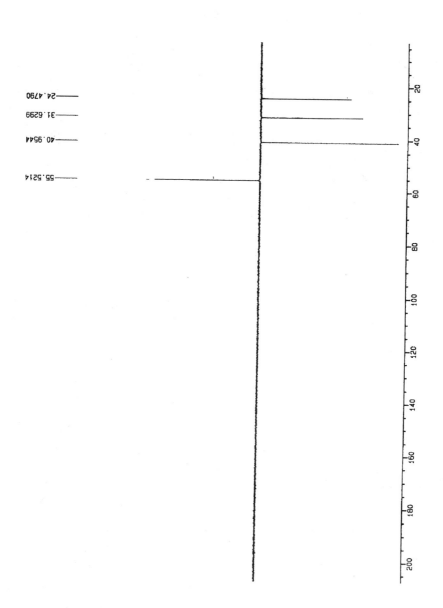

Table 3

Assignment for the Observed ^{13}C-NMR Bands of Arginine

Chemical Shift (ppm)	Assignment (Carbon #)
183.07	1
156.70	6
55.50	2
41.00	5
31.60	3
24.5	4

4.   Methods of Analysis

4.1   Compendial Tests

4.1.1   United States Pharmacopoeia

The USP contains a number of methods that define the compendial article:

		Specification
Identification	General method <197K>	Must conform
Specific Rotation	General method <781S>	NLT +26.3° and NMT +27.7° (Test solution: 80 mg per mL, in 6 N hydrochloric acid)
Loss on Drying	General method <731>	NMT 0.5% (dried at 105°C for 3 hours)
Residue on Ignition	General method <281>	NMT 0.3%
Chloride	General method <221>	NMT 0.05% (1.0 g shows no more chloride than corresponds to 0.70 mL of 0.020 N HCl)
Sulfate	General method <221>	NMT 0.03% (1.0 g shows no more sulfate than corresponds to 0.30 mL of 0.020 N H_2SO_4)
Iron	General method <241>	NMT 0.003%
Heavy Metals	General method <231>, Method I	NMT 0.0015%
Organic Volatile Impurities, Method I <467>:	General method <467>, Method I	Meets the requirements (using water as the solvent)
Assay	Titration	NLT 98.5% and NMT 101.5% (anhydrous basis)

4.1.2 European Pharmacopoeia

The EP contains a number of methods that define the compendial article:

		Specification
Identification	Primary: Tests A, C Secondary: A, B, D, E	A. Complies with specific optical rotation. B. solution is strongly alkaline. C. The Infrared absorption spectrum conforms. D. The principal spot is equivalent to that of the standard in the ninhydrin-positive substance test. E. Yields the expected reaction with β-naphthol and hypochlorite.
Appearance of Solution	General method (2.2.1)	Solution is clear, and less colored than reference solution BY_6.
Specific Optical Rotation	General method (2.2.7)	NLT 25.5° and 28.5°
Ninhydrin-Positive Substances	Thin-layer chromatography	Must conform
Chloride	General method (2.4.4)	NMT 200 ppm
Sulfate	General method (2.4.13)	NMT 300 ppm
Ammonium	Reaction with litmus paper	NMT 200 ppm
Iron	General method (2.4.9)	NMT 1 ppm
Heavy Metals	General method (2.4.8)	NMT 1 ppm
Loss on Drying	General method (2.2.32)	NMT 0.5% (dried at 100-105°C
Sulfated Ash	General method (2.4.14)	NMT 0.1%
Assay	Titration	NLT 98.5% and NMT 101.0% (anhydrous basis)

4.2   Elemental Analysis

Carbon	41.37 %
Hydrogen	7.05 %
Oxygen	18.39 %
Nitrogen	32.18 %

4.3   Titrimetric Analysis

The following procedure has been recommended for the titrimetric analysis of arginine [15]. Transfer about 80 mg of arginine (accurately weighed) to a 125 mL flask, dissolve in a mixture of 3 mL of formic acid and 50 mL of glacial acetic acid, and titrate with 0.1 N perchloric acid VS, determining the endpoint potentiometrically. A blank determination is performed, any any necessary corrections are to be made. Each milliliter of 0.1 N $HClO_4$ is equivalent to 8.710 mg of $C_6H_{14}N_4O_2$.

4.4   High Performance Liquid Chromatography

An application note describing the evaluation of a high-sensitivity amino acid analysis method for peptides and proteins has been described [16]. The method makes use of a combined OPA and a FMOC derivatization procedure and subsequent fluorescence detection. The AminoQuant system used in this work is based on a HP-1090 series II liquid chromatograph with a binary solvent pump. Rapid gas phase hydrolysis of 100-1000 pmol quantities of sample, and UV detection of amino acid derivatives, yielded accurate results for peptides and proteins. In this study, 30-300 pmol quantities of protein and peptide sample were subjected to rapid gas phase hydrolysis. The resulting hydrolyzates were analyzed using fluorescence detection of the amino acid derivatives. Arginine was tested in 5 proteins/peptides, and the overall accuracy for the various samples is as follows:

Urotensin	Ubiqutin	Lysozyme	Aspartate amino-transferase (AAT)	Saruplase
0.7	0.6	0.4	0.6	0.6

4.5   Determination in Body Fluids and Tissues

Several workers have published articles concerned with the determination of arginine in plasma. One of these reported an HPLC assay for the quantitation of L-arginine in human plasma [18]. The assay involves precolumn derivatization of arginine with naphthalene-dicarboxyaldehyde and cyanide, followed by HPLC using UV detection. The derivatized arginine was found to be stable, exhibiting less than 5% degradation in 20 hours. The calibration curve was generated in Ringer's lactate solution (instead of plasma) to correct for endogenous plasma L-arginine. The plasma recovery (relative to Ringer's solution for n = 4) was 103%. The mean intra-day assay precision (n = 6), expressed as coefficient of variation, was 3.4%, and the intra-assay precision (n = 6) was 7.0%. The methodology was applied to the quantitation of L-arginine in plasma samples from normal subjects who had been given a single oral (10 g) and a single intravenous dose (30 g) of exogenous L-arginine.

Another paper reported on the rapid analysis of nutritionally important free amino acids in serum and organs (liver, brain and heart) by liquid chromatography after precolumn derivatization with phenylisothiocyanate (PITC) [19]. This method was modified to include a change in column temperature (47.5°C compared to 25-35°C). By using a Waters Pico-Tag amino acid analysis (15 cm) column, separation of 27 PTC-amino acids in human serum and rat liver, brain or heart, was completed in 20 minutes. The total time for analysis and equilibration was 30 minutes, and the modified method was much faster than the traditional ion-exchange methods (2-3 hours).

Papers were cited reporting on the analysis of amino acid using dinitrophenylation and reverse-phase high-pressure liquid chromatography [20]. Others used state-of-the-art HPLC to analyze amino acids in physiological samples [21].

One paper reported the use of capillary gas-chromatographic determination of proteins and biological amino acids as the N(O)-tert-butyldimethylsilyl (tBOMSi) derivatives [22]. Forty seven biological amino acids were derivatized by a single-step reaction using N-methyl-N-(tert-butyldimethyl-silyl)trifluoroacetamide, and successfully separated on a HP-1 capillary column [22].

Another paper cited raised the problem of interference caused by nitro-L-arginine analogs in the *in vivo* and *in vitro* assay for nitrates and nitrites [23]. The effects of administration of nitro-containing and nitro-deficient L-arginine-derived nitric oxide synthase inhibitors on the measurement of nitric oxide in plasma, urine, and HEPES buffered physiological salt solutions was studied by ozone chemiluminescence and by the modified Griess reaction [23].

5.   Stability

Arginine is stable under ordinary conditions of use and storage, but is ordinarily protected from light. It may produce carbon monoxide, carbon dioxide, nitrogen oxides, and hydrogen chloride when heated to decomposition. Hazardous polymerization will not occur. The substance is known to be incompatible with strong oxidizers.

6.   Drug Metabolism and Pharmacokinetics

6.1   Metabolism

L-Arginine is metabolized by nitric oxide synthases (NOS) to nitric oxide and L-citrulline, or by arginase to urea and L-ornithine. L-ornithine is a precursor for polyamines that are required for cell proliferation and for proline, an essential component of collagen [14]. Many cells synthesize nitric oxide from the semi-essential amino acid, L-arginine, by virtue of NOS of constitutive forms (cNOS). These are expressed in healthy mediating vital functions, and inducible forms (iNOS), which are increasingly found in disease states [17]. Figure 9 shows the complete metabolism cycle of arginine and proline.

Figure 10 shows the biosynthesis of arginine in two different yeasts, *Candida* and *Sacchromyces*. The enzymes that play a role in the metabolic pathway are the arginases that are denoted as Arg 2 through 8 in the figure.

Figure 9.   Complete metabolism cycle of arginine and proline.

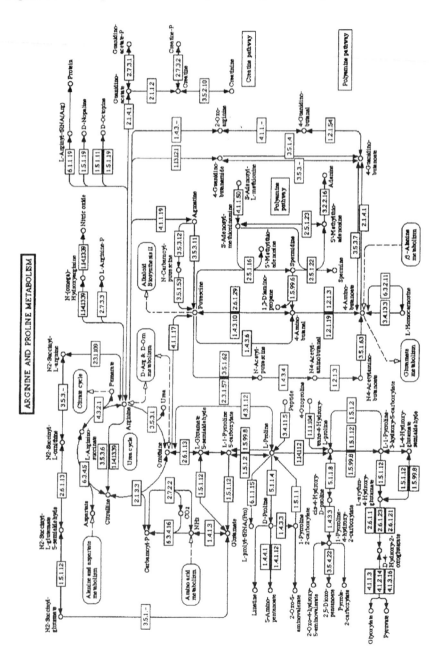

Figure 10.   Biosynthesis of arginine by *Candida* and *Sacchromyces*.

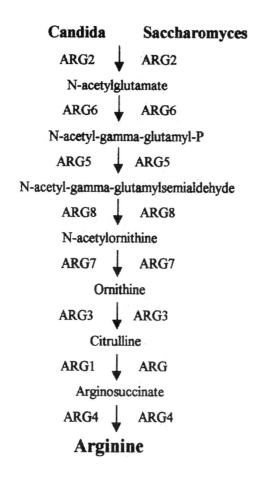

6.2   Pharmacokinetics and Pharmacodynamics

Several papers investigating the pharmacokinetics of arginine have been cited. One paper reports on the pharmacokinetics of L-arginine during chronic administration to patients with hypercholesterolaemia [24]. The study was designed to examine the disposition of L-arginine in hyper-cholesterolaemic subjects during long-term administration. Plasma L-arginine concentrations were determined by HPLC in 10 patients (eight women and two men, mean age 46 ± 16 years) after an intravenous dose of 10 or 30 g and an oral dose of 5 or 7 g. Pharmacokinetic studies were performed at regular intervals (4 weeks) during a 12-week period of oral administration of L-arginine (14-21 g/day). The average plasma L-arginine concentrations before (baseline) and during administration were found to be 16.1 ± 1.2 and 22.5 ± 1.3 µg/mL, respectively ($P < 0.05$). Plasma concentration of L-arginine remained above baseline throughout weeks 2-12. The L-arginine exposure, expressed as normalized area-under-the-curve for 8 hours after oral or intravenous doses during the first visit, was 894.4 ± 118.7 and 1837.8 ± 157.0 units respectively [24].

Another paper reported on the pharmacokinetic-pharmacodynamic relationship of L-arginine-induced vasodilation in healthy humans [25]. Pharmacokinetic studies were carried out after a single intravenous infusion of 6 g or 30 g, or after a single oral application of 6 g, as compared with the respective placebo in eight healthy male human subjects. L-arginine levels were determined by HPLC. The vasodilatation effect of L-arginine was assessed non-invasively by blood pressure monitoring and impedance cardiography. Urinary nitrate and cyclic GMP excretion rates were measured non-invasive indicators of endogenous NO production. Plasma L-arginine levels increased to (mean ± s.e. mean) 6223 ± 407 (range, 5100-7680) and 822 ± 59 (527-955) µmole/L after intravenous infusion of 6 and 30 g of arginine, respectively, and to 310 ± 152 (118-1219) µmole/L after oral injection of 6g arginine. Oral bioavailability of L-arginine was 68 ± 9 (5-87)%. Clearance was 544 ± 24 (440-620), 894 ± 164 (470-1190) and 1018 ± 230 (710-2130 mL/min, and elimination half-life was calculated as 41.6 ± 2.3 (34-55), 59.6 ± 9.1 (24-98) and 79.5 ± 9.3 (50-121) min, respectively, for 30g i.v., 6g i.v. and 6g p.o. of L-arginine. Blood pressure and total peripheral resistance were significantly decreased after intravenous infusion of 30g of L-arginine by $4.4 \pm 1.4\%$ and $10.4 \pm 3.6\%$, respectively, but were not significantly changed after oral or intravenous administration of 6g L-arginine [25].

Another paper reported on the pharmacokinetics of intravenous and oral L-arginine in normal volunteers [26]. This study was designed to examine the pharmacokinetics of single i.v. and oral doses of L-arginine in healthy volunteers (n = 10). A preliminary control study (n = 12) was performed to assess the variation in plasma L-arginine concentrations after ingesting a normal diet. The observed variation was taken into account when interpreting the data. The mean baseline plasma concentration of L-arginine in the control study was 15.1 ± 2.6 µg/mL. After intravenous administration (30 g over 30 minutes), the plasma concentration reached 1390 ± 596 µg/mL. The disappearance of L-arginine appeared biphasic, with an initial rapid disappearance due to concentration–dependent renal clearance, followed by a slower fall in plasma concentrations due to non-renal elimination. The peak concentration after oral administration (10g) was 50.0 ± 13.4 µg/mL occurring 1 hour after administration. Renal elimination was not observed after oral administration of this dose. The absolute bioavailability of a single oral 10g dose of L-arginine was approximately 20% [26].

6.3   Adverse Effects and Toxicity

Nausea, vomiting, flushing, headache, numbness, and local venous irritation may occur if arginine solutions are infused too rapidly. Elevated plasma potassium concentrations have been reported in uraemic patients, and arginine should therefore be administered with caution to patients with renal disease or amuria. Arginine hydrochloride should be administered cautiously to patients with electrolyte disturbances, as its high chloride content may lead to the development of hyperchoraemic acidosis [12].

Two alcoholic patients with severe liver disease and moderate renal insufficiency developed severe hyperkalaemia following administration of arginine hydrochloride, and one died. Both patients had received a total dose of 300 mg of spironolactone some time before arginine hydrochloride administration, but the contribution of spironolactone to the hyperkalaemia was not known [12].

Several papers investigating the toxicity of L-arginine have been reported. The first paper reported on the stimulation of lymphocyte natural cytotoxicity by L-arginine [27]. It was stated that *in vitro* L-arginine enhanced natural killer and lymphokine-activated-killer cell activity; with this cytotoxicity being mediated by CD56+ cells. *In vivo* arginine supplements (30 g/day for 3 days) increased the number of circulating CD56+ cells by a median of 32% in eight volunteers (P<0.01). This increase was associated with a mean rise of 91% in natural killer cell activity (P = 0.003) and of 58% in lymphokine-activated-killer cell activity (P = 0.001) in thirteen volunteers [27].

Another paper reported the use of high doses of dietary arginine during repletion impair weight gain and increased infectious mortality in protein-malnourished mice [28]. Protein malnutrition was induced by feeding mice for 6 weeks on an isoenergetic diet containing only 10 g protein/kg. Mice were then allowed to consume diets with normal amounts of protein (200 g/kg provided as amino acid mixtures of glycine and arginine in which arginine content ranged from 0 to 50 g/kg). During the repletion period, a significant weight gain was noted in the group fed on diets with either 10 or 20 g arginine/kg but not in the group fed on diet with 50 g arginine/kg, relative to the diet lacking arginine. Mortality rates after infection with *Salmonella typimurium* were not decreased by the addition of 10 or 20 g arginine/kg to the diet, and were in fact worsened by supplementation with 50 g arginine/kg. The result of this work showed that high doses of arginine become toxic. Mice fed on higher doses showed significant impairment of weight gain, and increased mortality rates [28].

Acknowledgement

The L-arginine used in this study was obtained from Pacific Pharmachem USA. The authors sincerely and appreciably thank Dr. Leroy Shervington for his valuable technical help, support, and advice throughout the work. The authors also wish to thank Dr. Ann Newman and Mr. Imre Vitez for providing the x-ray powder diffraction pattern and associated data.

References

1. J. Daly, *Surgery*, **112**, 56-67 (1992).

2. *The Life Extension Manual*, Colorado Futurescience, Inc., 1991-1999.

3. A. Barbul, G. Rettura, and S. M. Levenson, *Surgery Forum*, **28**, 101 (1980).

4. G. Rettura, A. Barbul, and S. M. Levenson, *J. Patent. Enter. Nutr.*, **1**, 22A (1977).

5. A. Barbul, H. L. Wasserkrug, and E. Seifter, *J. Surg. Res.*, **29**, 228 (1980).

6. E. Seifter, G. Rettura, and A. Barbul, *Surgery*, **84**, 224 (1978).

7. A. Barbul, S. A. Lazarou, D. Efron, H. L. Wasserkrug, and G. Efron, *Surgery*, **108**, 331-337 (1990).

8. A. Barbul, G. Rettura, and H. L. Wasserkrug, *Surgery*, **90**, 244-251 (1981).

9. A. Barbul, R. Fishel, S. Shimazu, H. L. Wasserkrug, N. Yoshimura, R. Tao, and G. Efron, *J. Surgical Research*, **38**, 328-334 (1985).

10. M. Hamon, B. Vallet, C. Bauters, N. Wernert, E. McFadden, J. Lablanche, B. Dupuis, and M. Bertrand, *Circulation*, **90**, 1357-1362 (1994).

11. S. Smith, M. Wheeler, H. Foster, and R. Weiss, *J. Urology*, **158**, 703-708 (1997).

12. *Martindale, The Extra Pharmacopoeia*, The Royal Pharmaceutical Society, Volume 31, 1996, pp. 1353-354.

13. *Merck Index*, 12th Edition, 1994, p. 132.

14. A. Lehninger, D. Nelson, and M. Cox, *Principle of Biochemistry"*, 2nd Edition, Worth Publisher, 1996.

15. *United States Pharmacopoeia 23*, United States Pharmacopoeial Convention, Rockville, MD, 1995, p., NF 18, page 129.

16. R. Grimm, Hewlett-Packard application Note, publication number 2-509-4585E (1992).

17. L. Kobzik, D.S. Bredt, C.J. Lowenstein, J. Drazen, B. Gaston, and D. Sugarbaker, *Am. J. Respir Cell Mol. Biol.*, **9**, 371-377 (1993).

18. V. Goplakrishnan, P. J. Burton, and T. F. Blaschke, *Anal. Chem.*, **68**, 3520-3523 (1996).

19. G. Sarwar and H. Botting, *J. Assoc. Off. Anal. Chem.*, **73**, 470-475 (1990).

20. R. Morton and G. Gerber, *Anal. Biochem.*, **170**, 220-227 (1988).

21. D. Fekkes, *J. Chrom. B, Biomed. Appl.*, **682**, 3-22 (1996).

22. K.L. Woo and D. S. Lee, *J. Chrom. B, Biomed. Appl.*, **665**, 15-25 (1995).

23. S.S. Greenberg, J. M. Xie, J. J. Spitzer, J. F. Wang, and J. Lancaster, *Life Science*, **57**, 1949-1961 (1995).

24. O. Tangphao, S. Chalon, H. Moreno, B. Hoffman, and T. Blaschke, *Clin. Sci.*, **96**, 199-207 (1999).

25. S. Bode-Boger, R. Boger, A. Galland, D. Tslkas, and J. Frolich, *Br. J. Pharmacol.*, **46**, 489-497 (1998).

26. O. Tangphao, M. Grossmann, S. Chalon, H. Moreno, B. Hoffman, and T. Blaschke, *Br. J. Clin. Pharmacol.*, **47**,26-266 (1999).

27. K.G. Park, P.D. Hayes, P. J. Garlick, H. Sewell, and O. Eremin, *Lancet*, **337**, 645-648 (1991).

28. M. D. Peck, *Br. J. Nutrition*, **75**, 787-795 (1995).

FENOTEROL HYDROBROMIDE

Abdulrahman A. Al-Majed

Department of Pharmaceutical Chemistry
College of Pharmacy
King Saud University
P.O. Box 2457
Riyadh-11451
Saudi Arabia

ANALYTICAL PROFILES OF
DRUG SUBSTANCES AND EXCIPIENTS
VOLUME 27

33

Contents

1.   Description [1,2]

1.1   Nomenclature

1.1.1   Chemical Names [3,4]

1-(3,5-Dihydroxyphenyl)-2-(4-hydroxy-α-methylphen-ethylamino)ethanol.

5-[1-hydroxy-2-[[2-(4-hydroxyphenyl)-1-methylethyl]-amino]-ethyl]-1,3-benzenediol.

3,5-dihydroxy-α-[[(p-hydroxy-α-methylphenethyl)amino]-methyl]benzylalcohol.

1-(3,5-dihydroxphenyl)-1-hydroxy-2-[(4-hydroxyphenyl)-isopropylamino]ethane.

1-(p-hydroxyphenyl)-2-[[β-hydroxy-β-(3',5'-dihydroxy-phenyl)] ethyl]aminopropane.

1.1.2   Nonproprietary Name

Fenoterol hydrobromide

1.1.3   Proprietary Names [1,3]

Berotec, Dosberotec, Partusisten, Arium.

1.2   Formulae

1.2.1   Empirical

Fenoterol $C_{17}H_{21}NO_4$

Fenoterol HBr $C_{17}H_{22}NO_4Br$

1.2.2   Structural

1.3   Molecular Weight

Fenoterol   303.4

Fenoterol HBr   384.3

1.4   Chemical Abstract System Registry Numbers [3]

Fenoterol   [13392-18-2]

Fenoterol HBr   [1944-12-3]

Fenoterol HCl   [1944-10-1]

1.5   Appearance and Color [3]

A white crystalline powder [3].

1.6   Uses and Applications

Fenoterol (*p*-hydroxyphenylorciprenaline) is a resorcinol derivative of metaproterenol (orciprenaline). It is a direct-acting sympathomimetic agent with a β-adrenergic activity and a selective action on β_2 receptors [1]. It is an effective bronchodilator with minimal cardiovascular adverse effects. In equipotent doses, fenoterol is comparable to albuterol and terbutaline, and clinically superior to isoproterenol, isoetharine, and metaproterenol. It is indicated in the treatment of bronchiospasm associated with asthma, bronchitis and other obstructive airway diseases [2].

2.   Methods of Preparation [6]

The synthesis of fenoterol starts with side chain bromination of *m*-diacetoxy-acetophenone (**I**), followed by displacement of halogen by 1-(*p*-methoxyphenyl)-2-*N*-benzylaminopropane to give (**II**). The benzyl group is then removed by hydrogenation. Hydrobromic acid is used to cleave the ether and the ester groups, and either catalytic or hydride reduction completes the synthesis of fenoterol according to the following scheme. Separation of the diastereomers was achieved by fractional crystallization.

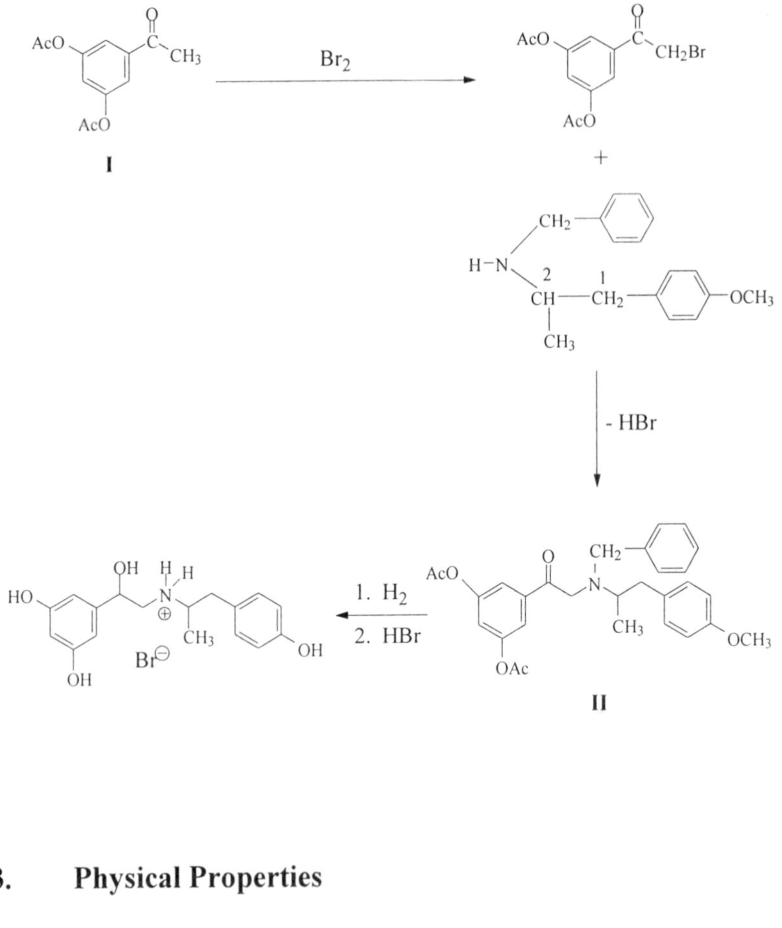

3.   Physical Properties

3.1   X-Ray Powder Diffraction Pattern

The x-ray powder diffraction pattern of fenoterol hydrobromide was performed using a Simons XRD-5000 diffractometer, and the powder pattern is shown in Figure 1). A summary of the crystallographic data deduced from the powder pattern of fenoterol hydrobromide is located in Table 1.

Figure 1

X-Ray Powder Diffraction Pattern of Fenoterol Hydrobromide

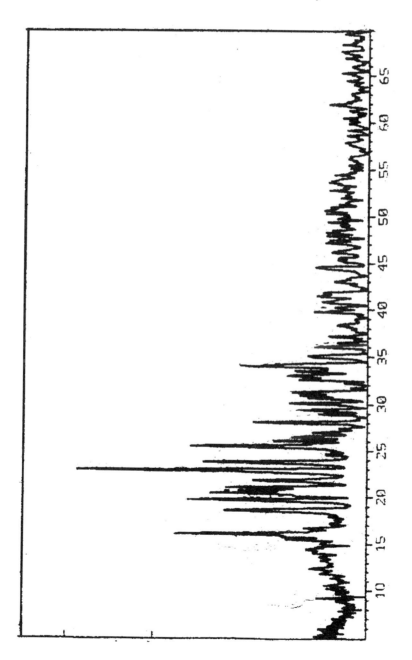

Table 1

Principal Lines Observed in the X-Ray Powder Diffraction of
Fenoterol Hydrobromide

Scattering Angle (degrees 2θ)	d-Spacing (Å)	Relative Intensity
9.221	9.5825	2.93
15.494	5.7142	7.73
16.091	5.5035	43.18
16.664	5.3156	4.68
18.585	4.7702	23.98
19.42	4.4933	37.95
20.200	4.3924	6.46
20.470	4.3351	28.54
20.756	4.2728	19.61
21.064	4.2141	23.56
21.837	4.0668	15.37
22.905	3.8794	100.00
23.784	3.7381	31.59
25.468	3.4945	36.59
26.076	3.4144	10.30
26.502	3.3605	8.11
26.944	3.3063	3.47
28.036	3.1800	15.30
28.938	3.0829	2.73
29.370	3.0385	5.18
30.090	2.9674	7.04
30.645	2.9150	4.05
30.924	2.8893	5.33
31.206	2.8638	7.71
32.544	2.7490	5.55
32.982	2.7135	8.09
33.417	2.6792	6.21
33.962	2.6375	17.99
34.200	2.6196	7.24
34.999	2.5616	4.17
36.017	2.4915	3.28
36.478	2.4611	1.25
37.106	2.4209	1.75

Table 1 (continued)

Scattering Angle (degrees 2θ)	d-Spacing (Å)	Relative Intensity
39.753	2.2656	3.32
40.223	2.2402	1.25
40.753	2.2123	2.37
41.745	2.1620	2.35
44.400	2.0386	2.54
46.158	1.9650	1.52
47.466	1.9138	1.70
47.873	1.8985	1.96
49.398	1.8434	1.89
59.132	1.5611	0.50
83.572	1.1560	0.78
22.905	3.8794	100.00
16.091	5.5035	43.18
19.42	4.4933	37.95
25.468	3.4945	36.59
23.784	3.7381	31.59
20.470	4.3351	28.54
18.585	4.7702	23.98
21.064	4.2141	23.56
20.756	4.2760	19.61
33.962	2.6375	17.99
21.837	4.0668	15.37
28.036	3.1800	15.30
26.076	3.4144	10.30
26.502	3.3605	8.11
32.982	2.7135	8.09
15.494	5.7142	7.73
31.206	2.8638	7.71
34.200	2.6196	7.24
30.090	2.9674	7.04
20.200	4.3924	6.46
33.417	2.6792	6.21
32.544	2.7490	5.55
30.924	2.8893	5.33
29.370	3.0385	5.18
16.664	5.3156	4.68

Table 1 (continued)

Scattering Angle (degrees 2θ)	d-Spacing (Å)	Relative Intensity
34.999	2.5616	4.17
30.645	2.9150	4.05
26.944	3.3063	3.47
39.754	2.2656	3.32
36.017	2.4915	3.28
9.221	9.5825	2.93
28.938	3.0829	2.73
44.400	2.0386	2.54
40.753	2.2123	2.37
41.745	2.1620	2.35
47.873	1.8985	1.96
49.398	1.8434	1.89
37.106	2.4209	1.75
47.466	1.9138	1.70
46.158	1.9650	1.52
40.223	2.2402	1.25
36.478	2.4611	1.05
83.572	1.1560	0.78
59.132	1.5611	0.50

3.2   Thermal Behavior

3.2.1   Melting Behavior

Fenoterol hydrobromide exhibits a melting point at 280°C, which is accompanied by decomposition [3].

3.2.2   Differential Scanning Calorimetry

The differential scanning calorimetry (DSC) thermogram of fenoterol hydrobromide was obtained using a Dupont TA-9900 thermal analyzer system, and the data processed by the Dupont data unit. The thermogram shown in Figure 2 was obtained at a heating rate of 10°C/min, and was run from 100 to 300°C. The compound was found to melt at 238.11°C, and the enthalpy of fusion was calculated to be 4.0 KJ/mole.

3.3   Solubility

Fenoterol hydrobromide is soluble in water and in 96% ethanol, but is practically insoluble in chloroform and in water [5].

3.4   Ionization Constants

Fenoterol hydrobromide is characterized by the existence of two ionization constants, for which $pKa_1 = 8.5$ and $pKa_2 = 10.0$ [3].

The pH value of 4% w/v solution has been reported to be in the range 4.5 to 5.2 [5].

3.5   Spectroscopy

3.5.1   UV/VIS Spectroscopy

The ultraviolet spectrum of fenoterol hydrobromide was recorded on a Shimadzu Ultraviolet-Visible (UV-VIS) 1601 PC Spectrophotometer. The UV measurement was carried out utilizing matched 1-cm quartz cells, and the substance dissolved in 96% ethanol.

Figure 2

Differential Scanning Calorimetry Thermogram of Fenoterol
Hydrobromide

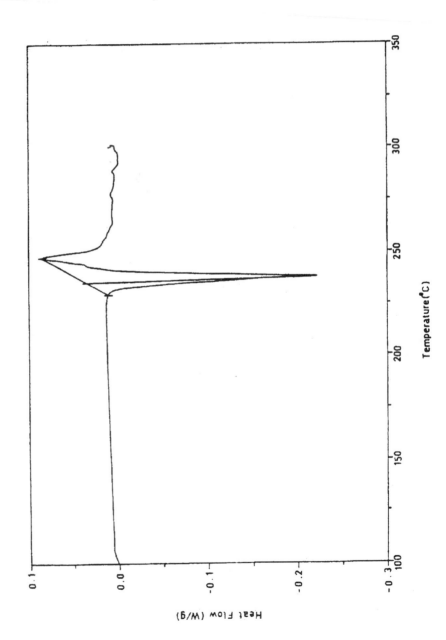

As shown in Figure 3, the spectrum consists of three absorption maxima. The first is located at 211 nm ($A_{1\%-1\,cm}$ = 561 and molar absorptivity = 2200 L/mole·cm), the second at 224 nm ($A_{1\%-1\,cm}$ = 446 and molar absorptivity 17124 L/mole·cm), and the third at 279 nm ($A_{1\%-1\,cm}$ = 101 and molar absorptivity = 3900 L/mole·cm). Clarke has also reported some absorption data [3]. In aqueous acid, the absorption maximum is observed at 275 nm, for which $A_{1\%-1\,cm}$ = 107. In aqueous alkali, the absorption maximum is observed at 295 nm, and $A_{1\%-1\,cm}$ = 190.

3.5.2 Fluorescence Spectrum

The native fluorescence spectrum of fenoterol hydrobromide in 96% ethanol was recorded using a Kontron Spectrofluorimeter, model SFM 25 A, equipped with a 150 W Xenon-High pressure lamp and driven by a PC Pentium-II computer. As shown in Figure 4, the excitation maximum was found to be 278 nm, and the emission maximum was located at 602 nm.

3.5.3 Vibrational Spectroscopy

The infrared spectrum of fenoterol hydrobromide was obtained using a Perkin-Elmer Infrared Spectrophotometer. The spectrum was obtained with the compound being compressed in a KBr pellet, and is shown in Figure 5. The principle absorption peaks were found at 690, 1150, 1320, 1500, 1590 and 1690 cm^{-1}.

Clarke reported the principal absorption bands at 700, 1160, 1200, 1510, 1575 and 1605 cm^{-1} [3].

3.5.4 Nuclear Magnetic Resonance (NMR) Spectra

The discussion of NMR assignments makes use of the following numbering scheme:

Figure 3

Absorption Spectrum of Fenoterol Hydrobromide

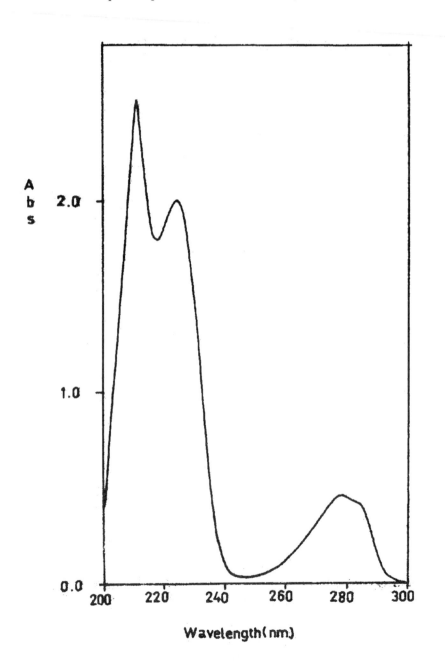

Figure 4

Fluorescence Spectrum of Fenoterol Hydrobromide

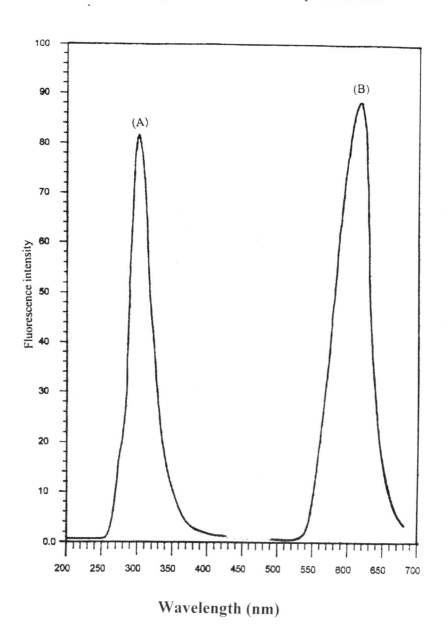

Figure 5

Infrared Spectrum of Fenoterol Hydrobromide

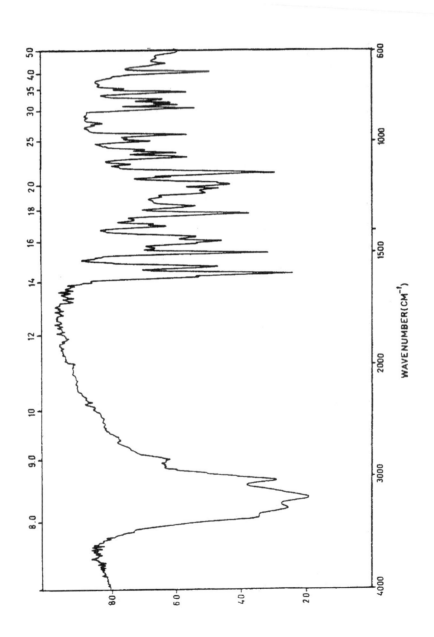

3.5.4.1 ¹H-NMR Spectrum

The proton NMR spectrum of fenoterol hydrobromide was obtained using a Bruker Advance Instrument operating at 300, 400, or 500 MHz. Standard Bruker software was used to obtain DEPT, COSY and HETCOR spectra. To obtain the solution-phase spectrum, the sample was dissolved in CH_3OH-d4, and tetramethylsilane (TMS) was used as the internal standard.

The ¹H-NMR spectrum of fenoterol hydrobromide is illustrated in Figure 6, and the spectral assignments deduced for the various protons are found in Table 2.

3.5.4.2 ¹³C-NMR Spectrum

The ¹³C-NMR spectrum of fenoterol hydrobromide was obtained using a Bruker Avance Instrument operating at 75, 100, or 125 MHz. To obtain the solution-phase spectrum, the sample was dissolved in CH_3OH-d4, and tetramethylsilane (TMS) was used as the internal standard.

The ¹³C-NMR spectrum of fenoterol hydrobromide is illustrated in Figure 7, and the spectral assignments deduced for the various carbons are found in Table 3.

3.5.5 Mass Spectrometry

The mass spectrum of fenoterol hydrobromide was obtained utilizing a Shimadzu PQ-5000 Mass Spectrometer, using helium as the carrier gas to obtain the parent ion. The detailed mass fragmentation pattern is shown in Figure 8. A base peak was observed at m/e = 44, along with a molecular ion peak at m/e = 384 along with other fragments. Table 4 contains the assignments made for the mass fragmentation pattern.

Figure 6

¹H-NMR Spectrum of Fenoterol Hydrobromide

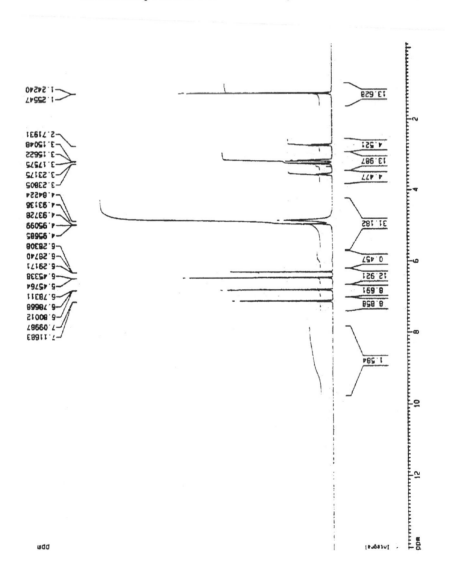

Table 2

Assignments for the Resonance Bands Observed in the ^{1}H-NMR
Spectrum of Fenoterol Hydrobromide

Chemical Shift, relative to TMS (ppm)	Number of protons	Multiplicity[*]	Proton Assignments
4.94	1H	dd	$H–C_1$
3.15	2H	m	$H–C_2$
3.54	1H	m	$H–C_3$
2.71, 3.24	2H	m, d	$H–C_4$
1.25	3H	d	$H–C_5$
6.45	2H	d	$H–C_2{}^a$
6.29	1H	dd	$H–C_4{}^a$
7.1	2H	d	$H–C_2{}^b$
6.79	2H	d	$H–C_3{}^b$
4.84	4H	------	–OH
4.84	1H	------	–NH

[*] d = doublet, m = multiplet, dd = doublet of doublet.

Figure 7

^{13}C-NMR Spectrum of Fenoterol Hydrobromide

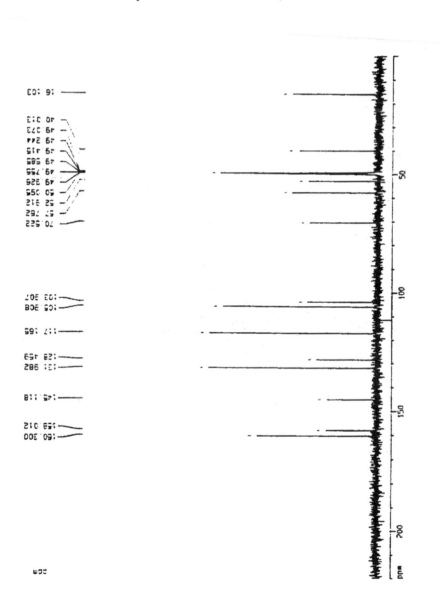

Table 3

Assignments for the Resonance Bands Observed in the ^{13}C-NMR
Spectrum of Fenoterol Hydrobromide

Chemical Shift, relative to TMS (ppm)	Multiplicity[*]	Assignment
70.52	d	C1
52.91	t	C2
57.76	d	C3
40.01	t	C4
16.10	q	C5
145.11	s	C1[a]
105.90	d	C2[a]
160.30	s	C3[a]
103.90	d	C4[a]
128.45	s	C1[b]
131.98	d	C2[b]
117.16	d	C3[b]
158.01	s	C4[b]

Figure 8

Mass Spectrum of Fenoterol Hydrobromide

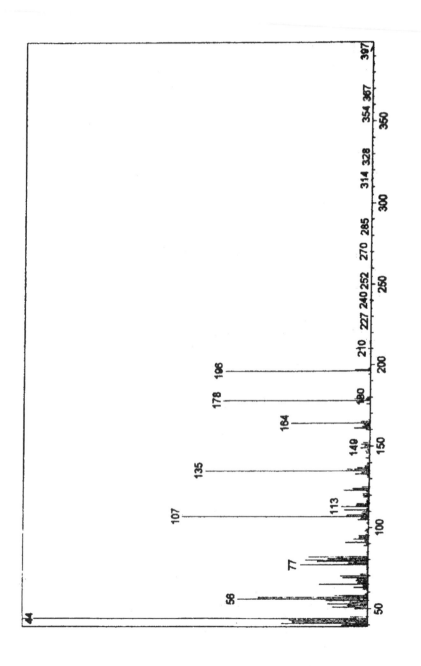

Table 3

Assignments for the Fragmentation Pattern Observed in the Mass
Spectrum of Fenoterol Hydrobromide

m/z	Relative intensity	Fragment
196	42%	
164	23%	
135	48%	
107	54%	
77	20%	
56	38%	$CH_2 \overset{\oplus}{=} N = CH - CH_3$
44	100%	$\overset{\oplus}{NH_2} = CH - CH_3$

4.   Methods of Analysis

4.1   Quantitative Official Methods

The British Pharmacopoeia [5] describes a titrimetric method for the determination of fenoterol hydrobromide as the pure drug substance. The method is performed by dissolving 0.6 g of the substance to be analyzed in water, and then adding 5 mL of 2M nitric acid, 25 mL of 0.1M silver nitrate, and 2 mL of ammonium iron (III) sulfate solution. The mixture is shaken, and titrated with 0.1M ammonium thiocyanate solution until the color becomes reddish yellow. The monographs requires the performance of a blank titration, and any necessary correction. Each milliliter of 0.1 M silver nitrate solution is equivalent to 38.43 mg of $C_{17}H_{21}NO_4 \cdot HBr$.

4.2   Identification

The British Pharmacopoeia [5] prescribes three identification tests:

(a)   When measured over the range of 230 to 350 nm, the absorption of a 0.01% w/v solution in 0.01M hydrochloric acid exhibits a maximum only at 275 nm and a shoulder at 280 nm. The absorbance at 275 nm is about 0.83 [5].

(b)   The infrared absorption is equivalent to the reference spectrum of fenoterol hydrobromide [5].

(c)   When tested according to the general method, the substance yields the reactions characteristic of bromides [5].

4.3   Elemental Analysis

The theoretical elemental composition of fenoterol is as follows [4]:

Carbon:	67.31%
Hydrogen:	6.98%
Nitrogen:	4.62%
Oxygen:	21.10%

4.4 Spectrophotometric Methods of Analysis

Abounassif and Abdel-Moety [7] determined fenoterol in tablets and in inhalation aerosols using two derivative-based spectrophotometric methods. In both methods, the absorbances of the solutions were measured at 295 nm and at 281 nm, respectively. Beer's law was obeyed for 10 to 70 μg/mL of fenoterol hydrobromide. The mean recoveries of the methods were in the range of 100.10 ± 0.29 % and 100.50 ± 0.76 for tablets, and 100.10 ± 0.28 % and 100.30 ± 0.60 for aerosols.

Tanabe *et al.* [8] described a colorimetric flow-injection method for the determination of fenoterol, orciprenaline, or terbutaline. The method involves reaction of the drugs with a 0.003% ethanolic solution of phenanthro[9, 10-d]imidazole-2-N-chloroimide in acidic media. The absorbance of the complex was measured at 530 nm, and calibration graphs were linear over the range of 0.5 to 200 μM for each drug substance. The recoveries exceeded 99%, and the precision (as measured by the coefficient of variation for n = 10) was better than 2%.

4.5 Chromatographic Methods of Analysis

4.5.1 Gas Chromatography

Couper and Drummer [9] determined fenoterol, terbutaline, and salbutamol in post-mortem blood by GC using orciprenaline as the internal standard. One milliliter of blood was mixed with 10 μL of an aqueous 10 μg/mL solution of the internal standard and 2 μL of 0.1 M sodium phosphate buffer (pH 7.5) in a polypropylene tube. After centrifugation, the supernatant was cleaned up on C_{18} Sep-Pak V_{ac} cartridge. The drugs were eluted from the cartridge with 1:1 methanol/acetonitrile, and subjected to GC analysis. The column (25 m x 0.2 mm i.d.) used was coated with Ultra-2, and operated with temperature programming from 90°C (held for 2 minutes) to 310°C (held for 15 minutes) at 15°C/min. Detection was by selected-ion monitoring EIMS.

4.5.2 Liquid Chromatography

Polettini *et al.* [10] reported the applicability of coupled-column liquid chromatography to the analysis of β-agonists in urine (fenoterol, clenbuterol, salbutamol, ractopamine, mabuterol isoprenaline, terbutaline,

and cimaterol) by direct sample injection. Urine containing the analyte
was filtered and injected onto a Microsphere C_{18} column (5 cm x 4.6 mm,
3 μm particle size). Elution was effected using 28.5% methanolic 0.1 M-
ammonium acetate in 0.01 M-triethylamine (pH 3.8), and detection was
made on the basis of the UV absorbance at 245 nm. After 5 minutes, the
sample fraction was automatically applied to a similar second column, and
eluted as before but with 35% methanol.

Doerge *et al.* [11] used liquid chromatography to determine β-agonist
residues (fenoterol, terbutaline, metaproterenol, salbutamol, and
clenbuterol) in human plasma. Human plasma containing the analytes was
diluted with water and applied to a Sep-Pak C_{18} cartridge. After washing
with water and acetonitrile, the analytes were eluted with methanol
containing 1% 100 mM-ammonium acetate (pH 4). The eluent was dried
under vacuum, and reconstituted in 77:23 acetonitrile/5.2 mM ammonium
acetate (pH 7). 20 μL Portions were separated on a 5 μm alpha-Chrom
C_{18} silica column (25 cm x 3 mm) combined with a 10 μm C_{18} silica guard
column (5 x 2 mm). Elution at a flow rate of 0.5 mL/min was performed
using a gradient method, over the range of 77-65% acetonitrile in
ammonium acetate. Detection was made on the basis of the UV
absorbance at 280 nm. Atmospheric-pressure CIMS was performed with
selected ion monitoring of MH^+ ions and low cone voltage conditions.

4.5.3   High Performance Liquid Chromatography

Jacobson and Peterson [12] described a HPLC method for the
simultaneous determination of fenoterol, ipratropium, salbutamol, and
terbutaline in nebulizer solutions. The method was performed on a Nova-
Pak C_{18} 4-μm Radial-Pak cartridge (10 cm x 8 mm) inside a compression
module, and detection was made at 220 nm. Ternary gradient elution was
initiated with 50% water and 50% aqueous 40% tetrahydrofuran (which
contained 2.5 mM-Pic B-8 Reagent Low UV). This solution was eluted
for up to 7.7 min, whereupon the solution composition was linearly
ramped over a 13-minute period to a medium consisting of 60% of the
original solution, 15% of water, and 25% of aqueous 50% methanol.

4.6   Capillary Electrophoresis-Mass Spectrometry Method

The determination of fenoterol, clenbuterol, and salbutamol by the online
coupling of capillary zone electrophoresis and mass spectrometry was

reported by Mazereeuw *et al.* [13]. The capillary (70 cm x 75 μm) was given a tapered tip by stretching it in a CH_4/O_2 flame. There was no electrode at the capillary outlet, but ground potential was supplied by the mass spectrometer inlet capillary at ground potential. The electric field between the capillary tip and the mass spectrometer created the electrospray. Samples were injected hydrodynamically, and electrophoresis was carried out at 30 kV.

4.7   Online ITP-CZE-ESP Method

Lamoree *et al.* [14] analyzed fenoterol, clenbuterol, terbutaline and salbutamol, by online isotachophoresis/capillary zone electrophoresis (ITP/CZE). The method used a single, untreated fused-silica capillary (74 cm x 100 μm), at a constant voltage of +20 kV. This was followed by electrospray mass spectrometry. Mass electropherograms were presented for a 870 nL test mixture containing 53 pg of clenbuterol, 56 pg each of terbutaline and fenoterol, and 59 pg of salbutamol, with detection limits being found to be in the 0.1 μM range. The use of the system as a trace analysis technique was demonstrated by the determination of these drugs in calf urine.

4.8   Online EE-ITP-CZE-ESP-MS Method

Van-der-Vlis *et al.* [15] developed a combined liquid-liquid electroextration (EE), isotachophoresis (ITP), capillary zone electrophoresis (CZE), electrospray (ESP), and mass spectrometry (MS) method for the determination of fenoterol, clenbuterol, salbutamol, and terbutaline. Sample focusing and separation by online EE-ITP-CZE was made using a fused-silica capillary (70 cm x 100 μm), coupled via a custom-made polyethylene connector to an untreated fused-silica capillary (20 cm x 100 μm) inserted directly into the stainless-steel needle assembly of the electrospray source of a Finnigan MAT TSQ-70 triple quadrupole mass spectrometer. Detection limits achieved by this method were 2 nM for clenbuterol, salbutamol, and terbutaline, and 1 nM for fenoterol.

4.9   Coulometric Method

Nikolic *et al.* [16] reported the use of a coulometric method for the analysis of fenoterol in the presence of other anti-asthmatic compounds. The method is based on the coulometric titration of the investigated compounds with electrogenerated chlorine in the presence of methyl orange indicator. The methanolic sample solutions of the drugs (0.1 mg/mL) were placed in the anode compartment of an electrolytic cell containing 0.5M $-H_2SO_4$, 0.2M NaCl, and methyl orange indicator solution. A constant current of 1 mA was passed through the solutions until the color was bleached, and the time taken for the titrant generation was measured against a blank with a chronometer. The concentration of the analyte was measured using the usual laws of Faraday. The presence of various ingredients and excipients in the investigated pharmaceutical preparation did not interfere with the electrochemical and chemical processes, which proceeded quantitatively.

4.10   Enzyme Immunoassay Method

Haasnoot *et al.* [17] developed an enzyme immunoassay method for the determination of fenoterol and ractopamine in urine, using antibodies against fenoterol-bovine serum albumin and fenoterol coupled to horseradish peroxidase (HRP). The calibration graph of fenoterol and ractopamine showed linearity over the concentration ranges 0.1–5 and 0.2–25 ng/mL, respectively.

4.11   Radioreceptor Assay Method

Helbo *et al.* [18] described a radioreceptor assay method for the estimation of fenoterol in the presence of other β_2-adrenergic agonists in urine. The method is based on the use of cell membranes and [3H]-dihydroalprenolol as a radiotracer. After termination of the binding reaction, the bound and free fractions were separated by filtration of the solution through glass fiber filters in a vacuum manifold. The radioactivity retained on the filter was measured by scintillation counting.

4.12 Radioimmunoassay Method

Rominger *et al.* [19] reported a radioimmunological determination method for fenoterol in biological samples. The mixture of antibodies against the enantiomers of fenoterol obtained with the selected hapten showed a high affinity for racemic fenoterol and for the ^{125}I-fenoterol used as a tracer. The limit of detection for fenoterol racemates in plasma and urine was 10-20 pg/mL.

4.13 Voltammetric Method

Boyd *et al.* [20] determined fenoterol in its dosage forms and in biological fluids by adsorptive stripping voltammetry using a Nafion-modified carbon paste electrode in conjunction with a platinum counter electrode and a Ag/AgCl reference electrode. Differential pulse stripping was carried out at 10 mV/s and with a pulse amplitude of 50 mV.

Boyd *et al.* [21] also estimated fenoterol, salbutamol, and metaproterenol using a similar method at unmodified and Nafion-modified-carbon paste electrodes.

5. Stability

Kobylinska-Luczko *et al.* [22] studied the stability of fenoterol hydrobromide in injection solutions and in tablets by a HPLC method. The influence of temperature, pH, and the choice of stabilizer on the stability of the drug forms was investigated. The amount of the active substance was controlled in the presence of degradation products. Lichrosorb RP-8 (particle size 10 µm) and a 35:65 mixture of methanol/0.1 M NaH_2PO_4 was used as the stationary and the mobile phases, respectively. Detection was carried out using a UV detector at 280 nm. The authors found that oxidation reactions exerted the most important influence on the stability of injection solutions. Fenoterol tablets were found to be more stable than were the injection solutions. In tablets stored for one year at room temperature, the decrease in the amount of active substance was less than 1%.

6.   Pharmacokinetics

Fenoterol is rapidly absorbed following oral ingestion or inhalation, and then conjugated primarily with sulfuric acid in man [2]. Administration of radioactive fenoterol indicated that in plasma, more than 90% of fenoterol exists as inactive metabolites [2,23]. Following oral and parenteral administration of the drug, the half-life is 7 hours, including both parent compound and metabolites [23,24]. A total of 60% of administered fenoterol was excreted in the urine within 24 hours, and total excretion was completed within 48 hours [24,25].

7.   Pharmacological Action

Fenoterol is a resorcinol derivative of metaproterenol, but with higher selectivity for β-2 adrenergic receptors [1]. It is comparatively more lipophilic than metaproterenol, and demonstrates enhanced β-2 selectivity and bronchodilation compared to metaproterenol [23]. The drug is a full β-2 adrenergic agonist that demonstrates the same degree of smooth muscle relaxation, but more potent bronchodilation if administered in equipotent doses as all previous marketed β-agonist agents [23,26-30]. Fenoterol, by virtue of its chemical structure that includes a large moiety attached to the terminal nitrogen, appears to have negligible α-adrenergic agonist activity, minimal β-1 adrenergic agonist activity, and selective β-2 adrenergic agonist activity [31-34].

8.   Therapeutic Doses

Fenoterol is active after inhalation or oral administration, and is indicated in the treatment of bronchiospasm associated with asthma, bronchitis, and other obstructive airway diseases [2]. In the treatment of bronchial asthma, fenoterol is administered by inhalation in 200-μg doses of 1 or 2 inhalations three times daily. It may also be given as a nebulized solution in doses of 0.5 to 2.5 mg inhaled up to 4 times daily [1]. The usual oral dose for the drug is 5 to 10 mg daily in 3 divided doses for 1 week [2].

Acknowledgements

The authors wish to thank Mr. Tanvir A. Butt, Department of Pharmaceutical Chemistry, College of Pharmacy, King Saud University, Riyadh, Saudi Arabia, for the typing of this profile.

References

1.   *Martindale, the Extra Pharmacopoeia*, 30th edn., J.E.F. Reynold, ed., The Royal Pharmaceutical Society, London (1993), pp. 1245-1246.

2.   R.C. Heel, R.N. Brogden, T.M. Speight, and G.S. Avery, *Drugs*, **15**, 3-32 (1978).

3.   *Clarke's Isolation and Identification of Drugs*, 2nd edn., A. C. Moffat, ed., The Pharmaceutical Press, London (1986), p. 616.

4.   *The Merck Index*, 12th edn., S. Budavari, ed., Merck and Co., NJ (1996), p.4022.

5.   *British Pharmacopoeia 1993*, Volume I, Her Majesty's Stationary Office, London, UK (1993), p. 274.

6.   D. Lednicer and L.A. Mitscher, *The Organic Chemistry of Drug Synthesis*, Wiley-Interscience, New York (1980), p. 38.

7.   M.A. Abounassif and E.A. Abdel-Moety, *Acta. Pharm. Jugosl.*, **39**, 359-363 (1989).

8.   S. Tanabe, T. Togawa, and K. Kawanabe, *Anal. Sci.*, **5**, 513-516 (1989).

9.   F.J. Couper and O.H. Drummer, *J. Chromatogr, Biomed. Appl.*, **685**, 265-272 (1996).

10.   A. Polettini, M. Montagna, E.A. Hogendoorn, E. Dijkman, P. Van-Zoonen, and L.A. Van-Ginkel, *J. Chromatogr.*, **695**, 19-31 (1995).

11.   D.R. Doerge, S. Bajic, L.R. Blankenship, S.W. Preece, and M.I. Churchwell. *J. Mass. Spectrom.*, **30**, 911-916 (1995).

12. G.A. Jacobson and G.M. Peterson, *J. Pharm. Biomed. Anal.*, **12**, 825-832 (1994).

13. M. Mazereeuw, A.J.P. Hofte, U.R. Tjaden, and J. Van-der-Greef, *Rapid Comm. Mass Spectrom.*, **11**, 981-986 (1997).

14. M.H. Lamoree, N.J. Reinhoud, U.R. Tjaden, W.M. A. Niessen, and J. Van-der-Greef, *Bio. Mass. Spectrom.*, **23**, 339-345 (1994).

15. E. van-der-Vlis, M. Mazereeuw, U.R. Tjaden, H. Irth, and J. Van-der-Greef, *J. Chromatogr.*, **712**, 227-234 (1995).

16. K. Nikolic, L. Arsenijevec, and M. Bogavac, *J. Pharm. Biomed. Anal.*, **11**, 207-210 (1993).

17. W. Haasnoot, P. Stouten, A. Lommen, G. Cazemier, D. Hooijerink, and R. Schilt, *Analyst*, **119**, 2675-2680 (1994).

18. V. Helbo, M. Vandenbroeck, and G. Maghuin-Rogister, *Arch. Lebensmittelhyg.*, **45**, 57-61 (1994).

19. K.L. Rominger, A. Mentrup, and M. Stiasni, *Arzneim. Forsch.*, **40**, 887-895 (1990).

20. D. Boyd, J.R. Barreira-Rodriguez, P. Tunon-Blanco, and M.R. Smyth, *J. Pharm. Biomed. Anal.*, **12**, 1069-1074 (1994).

21. D. Boyd, J.R. Barreira-Rodriguez, A.J. Miranda-Ordieres, P. Tunon-Blanco, and M.R. Smyth, *Analyst*, **119**, 1979-1984 (1994).

22. A. Kobylinska-Luczko, A. Gyzeszkiewics, I. Cendrowska, and K. Butkiewicz, *Chromatogr.*, **87**, 285-293 (1988).

23. N Svedmyr, *Pharmacotherapy*, **5**, 109-126 (1985).

24. K.L. Rominger and W. Pollman, *Arzneim. Forsch.*, **22**, 1196-1201 (1972).

25. S. Larsson and N. Svdmyre, *Ann. Allergy*, **39**, 362-366 (1977).

26. H.W. Kelly, *Clin. Pharm.*, **4**, 393-403 (1985).

27. H. Minatoya, *J. Pharmacol. Exp. Therap.*, **206**, 515-527 (1978).

28.   L. Nyberg and C. Wood, *Eur. J. Respir Dis.*, **65**, 1-290 (1984).

29.   C.G. Lofdahl, K. Svedmyr, N. Svedmyr, et al., *Eur. J. Respir. Dis.*, **65**, 124-127 (1984).

30.   N. Svedmyr and B. G. Simonsson, *Pharmacol. Therap.*, **3**, 397-440 (1978).

31.   S.R. O'Donnell, *Eur. J. Pharmacol.*, **12**, 35-43 (1970).

32.   H. Kundig, *Med. Proceedings*, **18**, 9 (1972).

33.   E. Cohen, A. As Van, and C. Quirk, *Med. Proceedings*, **18**, 24 (1972).

34.   R.T. Brittain, C.M. Dean, and D. Jack, "Sympathomimetic Bronchodilating", in ***Respiratory Pharmacology***, J. Widdicombe, ed., Pergamon Press, Oxford (1981) pp. 613-652.

FLUCONAZOLE

Alekha K. Dash[1] and William F. Elmquist[2]

(1) Department of Pharmaceutical & Administrative Sciences
School of Pharmacy and Allied Health Professions
Creighton University
Omaha, NE 68178
USA

(2) Department of Pharmaceutical Sciences
College of Pharmacy
University of Nebraska Medical Center
Omaha, NE 68198
USA

Contents

4.   **Methods of Analysis**
         4.1   Compendial Tests
         4.2   Elemental Analysis
         4.3   Spectrophotometric Methods of Analysis
         4.4   Gas Chromatography
         4.5   High Performance Liquid Chromatography
         4.6   Determination in Body Fluids and Tissues

5.   **Stability**
         5.1   Solution-Phase Stability
         5.2   Stability in Formulations
         5.3   Incompatibilities with Functional Groups

6.   **Pharmacokinetics and Metabolism**
         6.1   Animal Studies
         6.2   Normal Volunteer Studies
         6.3   Patient Studies

7.   **Toxicity**

8.   **Drug Interactions**

Acknowledgement

References

1.   Description

1.1   Nomenclature

1.1.1   Chemical Name

α-(2,4-Difluorophenyl)-α-(1-H-1,2,4-triazol-1-ylmethyl)-1*H*-1,2,4-triazole-1-ethanol

2,4-difluro-α, -α-bis (1H-1,2,4-triazole-1-ylmethyl)benzyl alcohol

2-(2,4-difluorophenyl)-1,3-bis (1*H*-1,2,4-triazol-1-yl)-propane-2-ol

1.1.2   Nonproprietary Names

Fluconazole

1.1.3   Proprietary Names

Diflucan

1.2   Formulae

1.2.1   Empirical

$C_{13}H_{12}F_2N_6O$

1.2.2   Structural

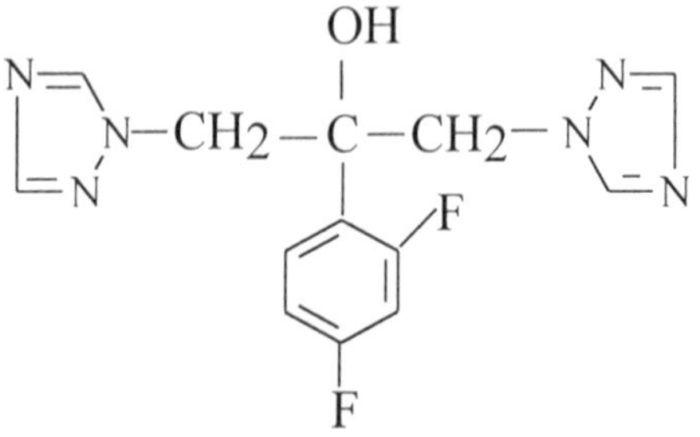

1.3   Molecular Weight

306.27

1.4   CAS Number

86386-73-4

1.5   Appearance

Fluconazole is obtained as a white crystalline powder [1].

1.6   History and Therapeutic Properties

Fluconazole is a synthetic antifungal agent belonging to the group of triazoles. The discovery of fluconazole (originally known as UK-49,858) is credited to a group of scientists led by Ken Richardson at Pfizer Central Research in Sandwich, Kent (UK) in 1981 [2]. The drug was approved by the FDA for use in the United States on January 9, 1990. The discovery team of this drug received numerous awards for their outstanding discovery of the world's leading antifungal drug including the Queen's Award for Science and Technology 1991, and the Discoverers Award of the PhRMA in 1994. This drug is highly effective against a variety of fungal pathogens that lead to systemic mycoses.

This drug is structurally related to the antifungal agents that are imidazole-derivatives. However, fluconazole differs markedly from other imidazoles in its pharmacokinetic properties. This agent is less lipophilic and more hydrophilic when compared to other azole antifungal agents. The presence of two triazole rings (bis-triazole) makes this compound less lipophilic and protein bound. The presence of a halogenated phenyl ring increases its antifungal activity. It acts as a fungistatic agent. Although the exact mechanism of action of fluconazole is not yet known, presumably it alters cellular membranes resulting in increased membrane permeability and thereby impaired uptake of precursor and leakage of essential elements from the cell. It is also believed that this drug inhibits cytochrome P-450 14-α-demethylase in fungi, which causes an accumulation of C-14 methylated sterols and decreases the concentration of ergosterol [3].

Fluconazole is available in three different dosage forms [4].
>   Tablets (50, 100, and 200 mg)
>   Suspensions (50 mg / 5 mL and 200 mg / 5 mL)
>   Parenterals (2 mg/mL) in 0.9% saline or in 5.6% dextrose.

2.   Methods of Preparation

The rational design, synthesis, and structure-activity relationships of fluconazole have been described by Richardson *et al.* [5,6]. The need for a safe and effective (both orally and intravenously) treatment for systemic fungal infection served to initiate this project in 1970. The rational design and synthesis of this compound was achieved in three major steps.

2.1   Replacement of imidazole by triazole

Initial investigations focused on imidazole derivatives. The two major problems encountered during the use of imidazole as an antifungal agent are (1) rapid first-pass metabolism, and (2) high levels of protein binding. A series of derivatives of imidazole, structurally similar to ketoconazole, were first synthesized (tetrahydrofurans, dioxolones, dithiololanes, and tertiary alcohols) as shown in Scheme 1. The tertiary alcohol series showed more activity in animal infection models. A series of compounds was then synthesized as shown in scheme 2, where imidazole group was replaced with other groups. Only the 1,2,4-triazole group showed promising results. Therefore, attention was geared to the development of triazole tertiary alcohol derivatives.

2.2   Addition of a second triazole group to the structure

The first metabolically stable compound developed was the 1,2,4-triazole. This compound (UK-47, 265) showed an excellent *in vivo* activity in a systemic candidosis model. However, this drug was found to be highly hepatotoxic. Finally, a series of bis-triazoles were synthesized in which the dichlorophenyl unit was replaced with range of groups. Two principal approaches were utilized for this synthesis, and these are shown in Schemes 3a and 3b. The chloroacetyl derivative was reacted with 1,2,4-triazole derivative to yield a ketone, which was then converted to an epoxide. Opening of the epoxide with 1,2,4 triazole produced the bis-triazole derivative (Scheme 3a). With the second approach, dichloroacetone was converted to the dichloropropanol derivative. This was then reacted with 1,2,4-triazole as shown in Scheme 3b.

Scheme 1.   Synthesis of structural types of imidazole derivatives.

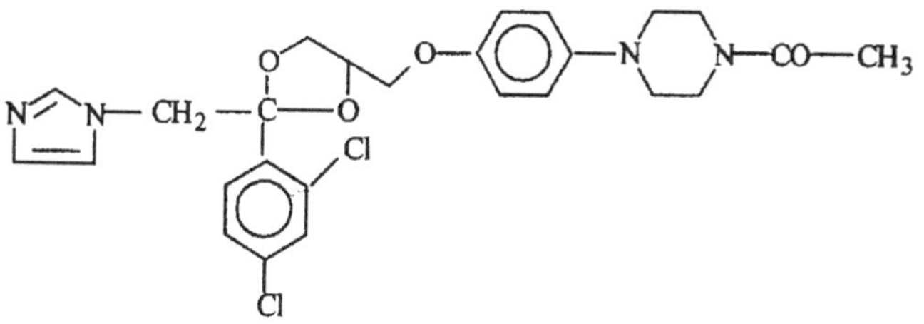

KETOCONAZOLE

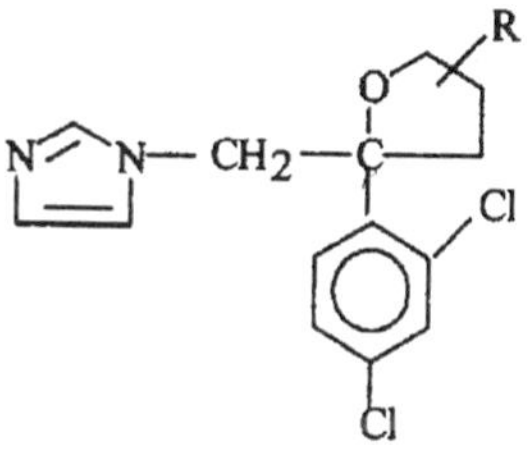

Tetrahydrofurans

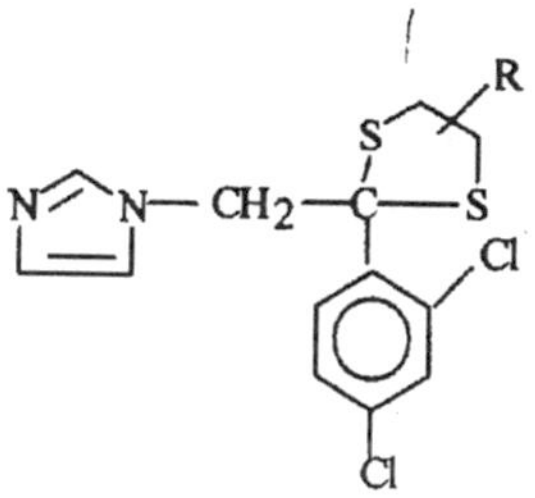

Dithiolanes

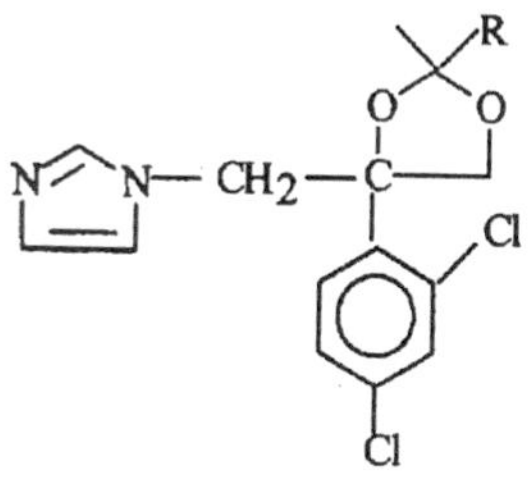

Dioxolanes

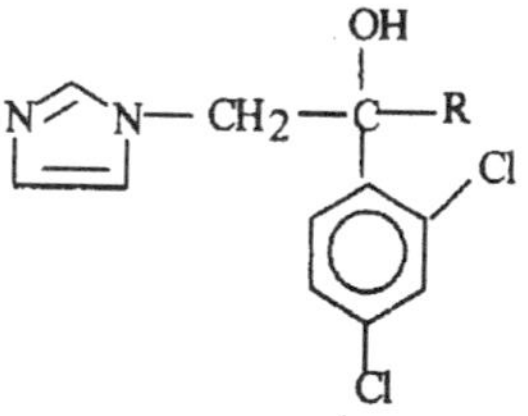

Tertiary Alcohols

Scheme 2. Synthesis of non-imidazole derivatives.

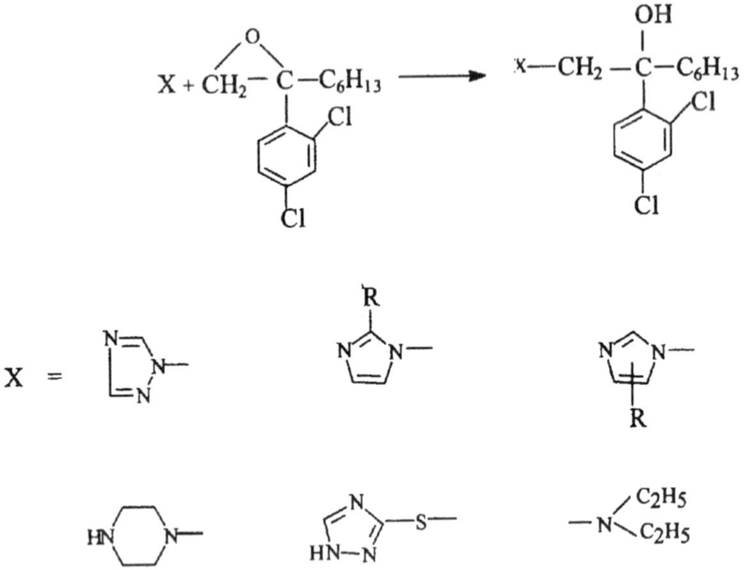

$$X + CH_2 - \overset{\displaystyle O}{\overset{\displaystyle \diagup\!\!\!\!\diagdown}{C}} - C_6H_{13} \longrightarrow X - CH_2 - \overset{\displaystyle OH}{\underset{|}{C}} - C_6H_{13}$$

X =

Scheme 3.   Synthesis of bis-triazoylpropane-2-ol derivatives.

(a) $Cl-CH_2-\overset{\overset{O}{\|}}{C}-R$ ⟶ triazole$-CH_2-\overset{\overset{O}{\|}}{C}\diagdown R$ ⟶ triazole$-CH_2-\overset{O}{C}-CH_2$ with R

bis-triazole $-CH_2-\overset{\overset{OH}{|}}{\underset{R}{C}}-CH_2-$ bis-triazole

(b) $Cl-CH_2-\overset{\overset{O}{\|}}{C}-CH_2-Cl$ ⟶ $Cl-CH_2-\overset{\overset{OH}{|}}{\underset{R}{C}}-CH_2-Cl$

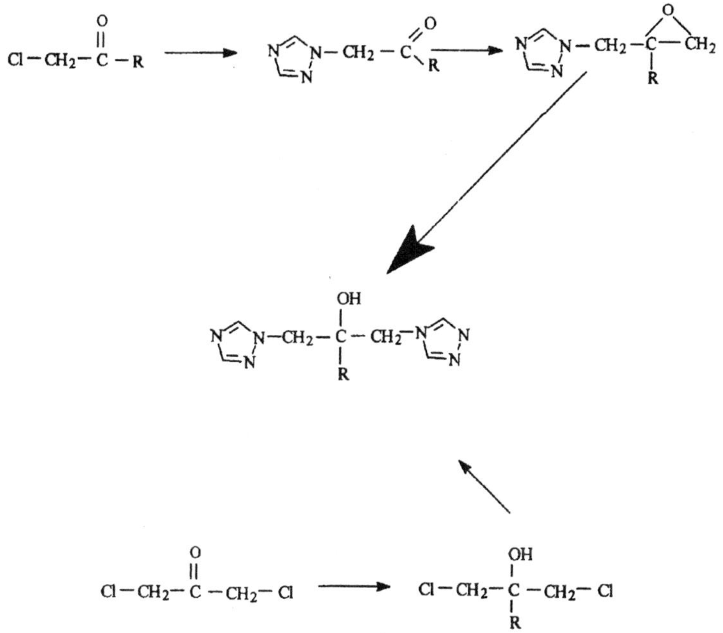

2.3   Substitution of difluorophenyl for dichlorophenyl

The last part of the synthesis included the substitution of 2-phenyl moiety with one or more halogen groups (Scheme 4). Substitution of the difluorophenyl for dichlorophenyl further decreased the lipophilicity (log P of 0.5 as opposed to 1.0) and increased the aqueous solubility of the compound (8-10 mg/mL). This compound, originally known as UK-49,858, was named Diflucan and given the generic name fluconazole.

Scheme 4.   Substitution of difluorophenyl for dichlorophenyl

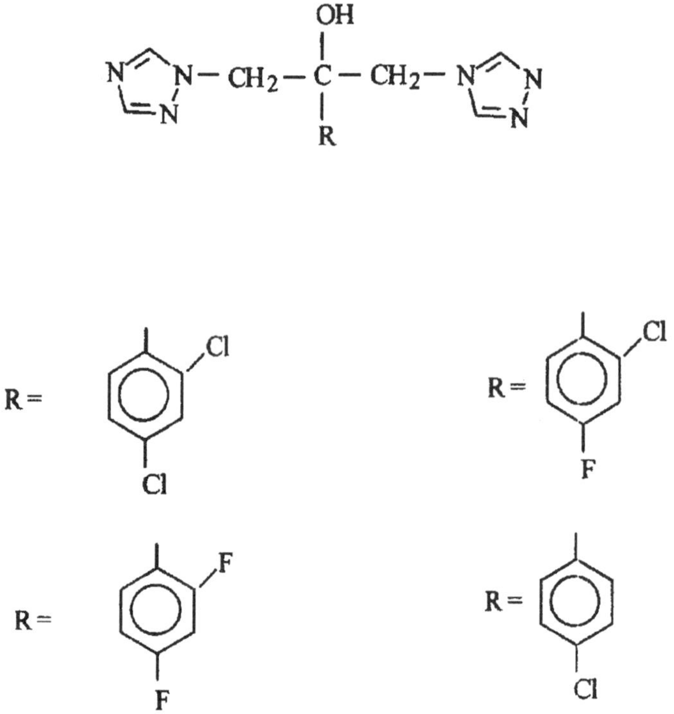

3. Physical Properties

3.1 X-Ray Powder Diffraction

X-ray powder diffraction patterns for various crystal forms of fluconazole were obtained using a wide-angle x-ray diffractometer (model D500, Siemens). The powder diffraction pattern of commercially available fluconazole is shown in Figure 1, and the crystallographic information calculated from the diffraction pattern is provided in Table 1 [7].

Scientists from Pfizer, Inc., have reported the existence of three polymorphic forms of fluconazole, which they have designated as forms I, II, III, and a hydrate [8]. The powder patterns for these forms are shown in Figures 2-5, respectively, and it appears that the commercially available form of the drug substance corresponds to form III.

3.2 Thermal Methods of analysis

3.2.1 Melting Behavior

The melting ranges of the three known polymorphs of fluconazole have been reported by Pfizer to fall within a very narrow temperature range [8]. Polymorphic Form I melts over the range of 135 – 136°C, Form II in the range of 138 – 140°C, and Form III melts in the range of 137 – 138°C.

3.2.2 Differential Scanning Calorimetry and Thermogravimetric Analysis

The differential scanning calorimetry thermogram of commercially available fluconazole was obtained using a Shimadzu model DSC-50. The sample was heated in a non-hermetically crimped aluminum pan at a rate of 10°C/min, over the range of 30-200°C. Thermogravimetric analysis of the sample was carried out using a Shimadzu model TGA-50 thermo-gravimetric analyzer.

The DSC and TG thermograms are shown in Figure 6 [7]. The melting endotherm was observed to yield a maximum at a temperature of 140.1°C, for which the enthalpy of fusion of was calculated to be 75 J/g. The anhydrous nature of the compound was evident in the TG thermogram, where only a 0.6% weight loss was noted over the temperature range of 115-200°C. This slight loss of mass is probably due to a small degree of sample decomposition.

Figure 1.   X-ray powder diffraction pattern of commercially available fluconazole.

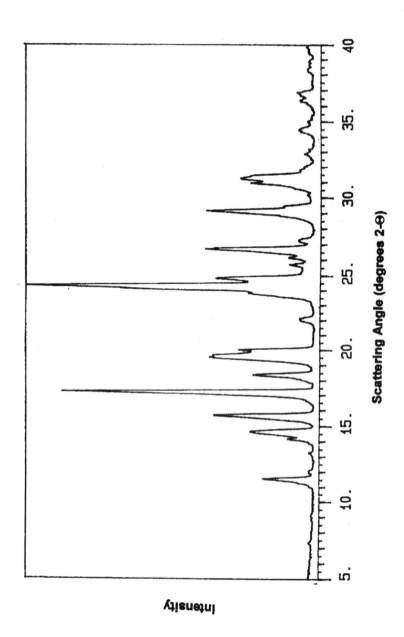

Table 1

Crystallographic Data from the X-Ray Powder Diffraction
Pattern of Fluconazole

Scattering angle (degrees 2-θ)	d-spacing (Å)	Relative Intensity
11.55	7.65	18.87
14.65	6.04	24.46
15.70	5.64	38.02
17.25	5.14	81.15
18.40	4.82	23.11
19.55	4.53	39.53
20.05	4.42	28.67
23.80	3.74	22.37
24.10	3.69	50.96
24.35	3.65	100.0
24.90	3.57	35.85
26.75	3.33	41.09
29.15	3.06	38.31
29.50	3.02	10.86
30.95	2.88	21.03
31.35	2.85	26.26

Figure 2.   X-ray powder diffraction pattern of fluconazole, Form I.

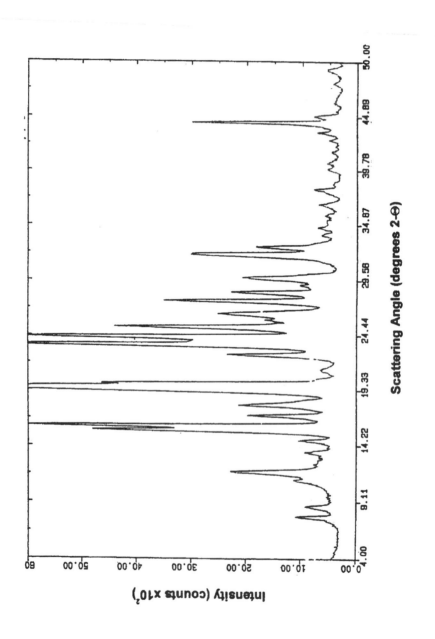

Figure 3.   X-ray powder diffraction pattern of fluconazole, Form II.

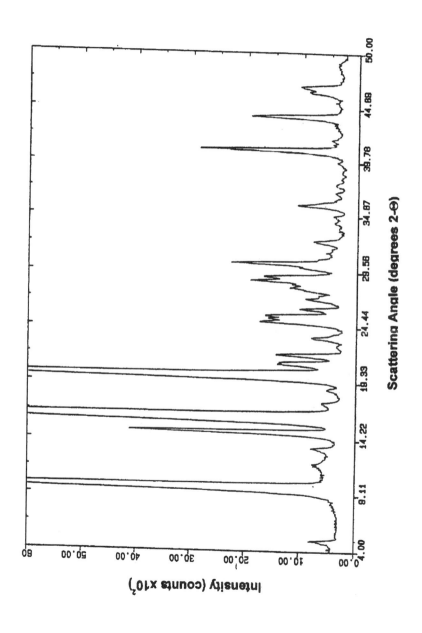

Figure 4.   X-ray powder diffraction pattern of fluconazole, Form III.

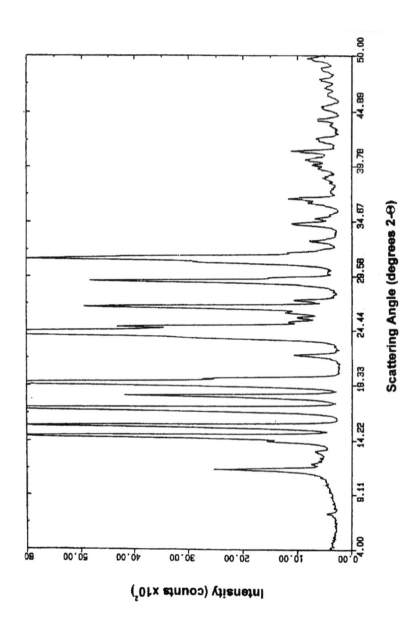

Figure 5.   X-ray powder diffraction pattern of fluconazole, monohydrate phase.

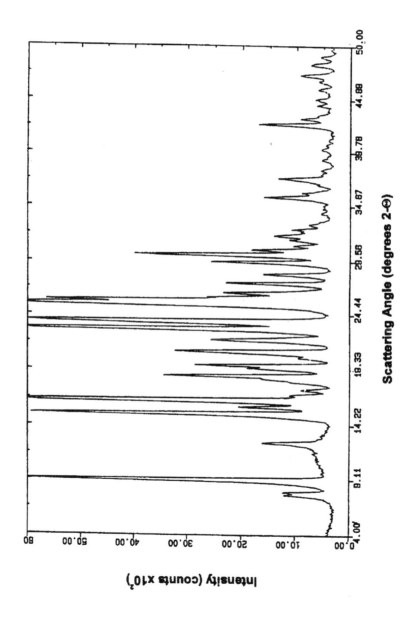

Figure 6.   Differential scanning calorimetry (curve **a**) and thermogravimetric analysis (curve **b**) thermograms of
commercially available fluconazole.

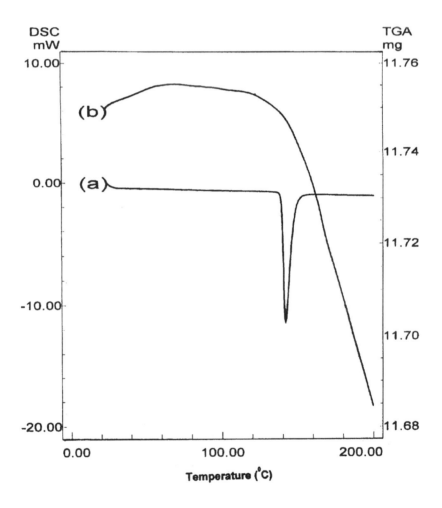

3.3   Solubility Characteristics

Solubility data at 23°C were obtained for fluconazole, Form III, using an ultraviolet absorption assay method to determine the concentration of substance present in saturated solutions [8]. These results are found in Table 2.

The aqueous solubility of fluconazole at 37°C has been reported to be 8 mg/mL [4].

The pH-solubility profile of fluconazole, Form II, was obtained at 4, 26 and 37°C, and is shown in Figure 7 [8]. These curves may be contrasted with those obtained for Form III at 4, 23, and 37°C and which are shown in Figure 8 [8].

The intrinsic dissolution profiles of the various forms of fluconazole (obtained from compressed disks) in water is shown in Figure 9 [8].

3.4   Partition Coefficients

The partition coefficient between octanol and water (reported as logP) of fluconazole was determined by the method of Stopper and McClean [9]. This study reported a 0.5 mg/mL of the total drug distributed between equal volumes (1 mL each) of octanol and buffer (0.1M sodium hydrogen phosphate, pH 7.4). This finding indicates that fluconazole has a logP equal to 0.5, which may be interpreted as signifying moderate lipophilicity. This value is quite low with respect to the logP values of most other azole antifungal agents (logP for ketoconazole is 3.5, and log P is 5.7 for itraconazole).

3.5   Ionization Constants

The ionization constant (pKa) of fluconazole was determined by solubility measurements, and found at 24°C in 0.1M NaCl solution to be 1.76 ± 0.10 [8]. Fluconazole is a weak base, and undergoes protonation at the N-4 nitrogen (see Scheme 5). The predominate protonation of the N-4 nitrogen of the 1H-1,2,4-triazole moiety has also been confirmed by ^{15}N NMR spectroscopy. Due to the weak basicity of this compound, secondary protonation is considered less likely in an *in vivo* situation.

Table 2

Solubilities of Fluconazole in Different Solvent Systems

Solvent	Solubility at 23°C (% w/v)	USP Definition
Water	0.5	Slightly soluble
Aqueous 0.1M HCl	1.4	Sparingly soluble
Aqueous 0.1M NaOH	0.5	Slightly soluble
Chloroform	3.1	Sparingly soluble
Acetone	4.0	Soluble
Propan-2-ol	0.8	Slightly soluble
Methanol	25.0	Freely Soluble
Ethanol	2.5	Sparingly soluble
n-Hexane	less than 0.1	Very slightly Soluble
Methanolic 0.01M HCl	30.5	Freely Soluble

Figure 7.   pH-solubility profiles of fluconazole, Form II.

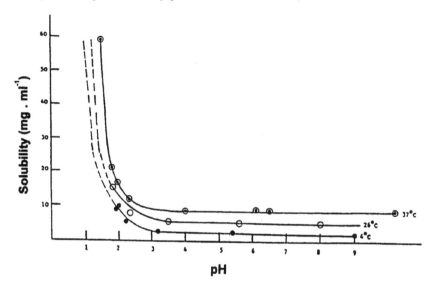

Figure 8.   pH-solubility profiles of fluconazole, Form III.

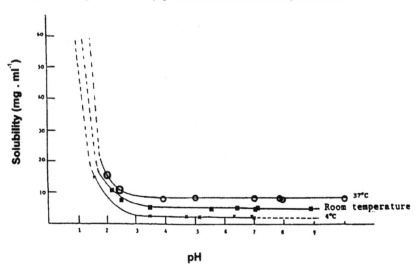

Figure 9.   Intrinsic dissolution profiles in water of fluconazole Forms
             I, II, III, and the monohydrate from compressed discs.

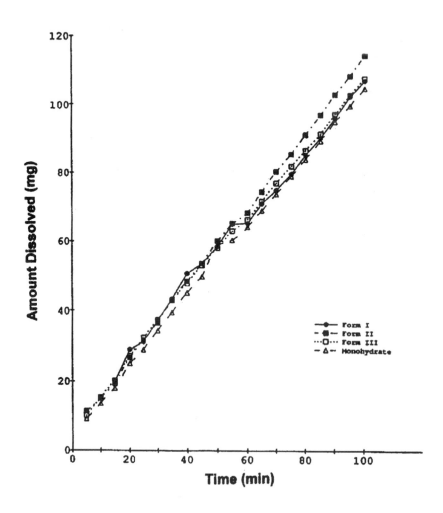

3.6   Spectroscopy

3.6.1   UV/VIS Spectroscopy

Ultraviolet absorption spectra of fluconazole were obtained using an model UV-1610 PC UV spectrophotometer (Shimadzu). Spectra were obtained in both methanol and in water at a solute concentration of 200 µg/mL, and are shown in Figures 10 and 11, respectively. Various peak maxima, their corresponding wavelengths and absorbance values, are shown in Table 3 [7].

3.6.2   Vibrational Spectroscopy

The infrared spectrum of fluconazole, Form III, was obtained using on a Perkin-Elmer model 983 spectrometer. Spectra for the compound as obtained in a potassium bromide disc and in a Nujol mull are shown in Figures 12 and 13, respectively. Assignments of the characteristic bands in the spectrum are listed in Table 4 [8].

3.6.3   Nuclear Magnetic Resonance Spectrometry

3.6.3.1 ^{1}H-NMR Spectrum

Proton nuclear magnetic resonance spectra of fluconazole were obtained using a Bruker WM-250 NMR spectrometer. Spectra of the compound dissolved in deuterated dimethyl sulfoxide (DMSO-D_6) and in deuterated dimethyl sulfoxide (DMSO-D_6) containing D_2O are shown in Figures 14 and 15, respectively. The chemical shifts, multiplicities, assignments, and effect of deuterium oxide addition on the chemical shifts are summarized in Table 5 [8]. The numbering scheme used in the table is derived from the following structure:

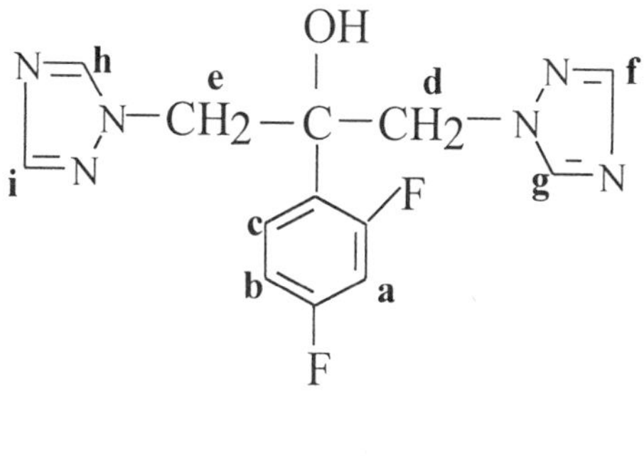

Table 3

Spectral Properties of Fluconazole in Different Solvent Systems

Peak Number	Wavelength (nm)	Absorbance
Methanol		
1	266.0	0.369
2	261.2	0.407
3	211.0	2.817
4	195.4	0.094
Water		
1	265.6	0.362
2	261.0	0.404
3	210.2	3.138

Figure 10. UV absorption spectrum of fluconazole, 200 µg/mL in methanol.

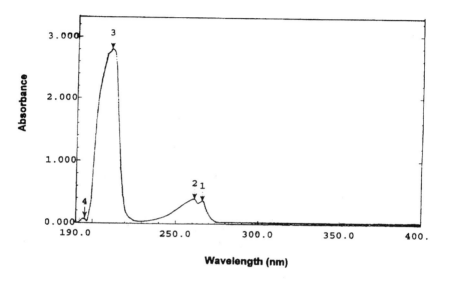

Figure 11. UV absorption spectrum of fluconazole, 200 µg/mL in water.

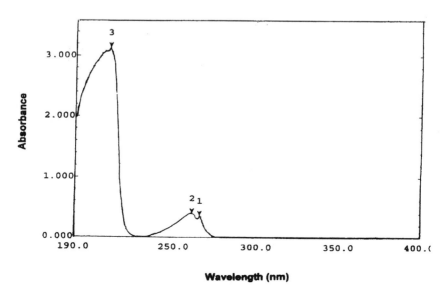

Figure 12.   Infrared absorption spectrum of fluconazole as a potassium
               bromide disk.

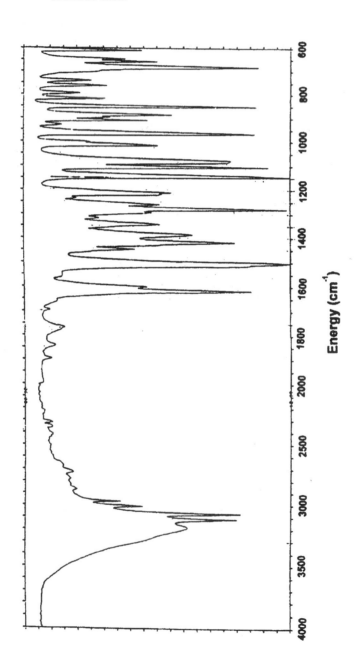

Figure 13.   Infrared absorption spectrum of fluconazole as a Nujol
mull.

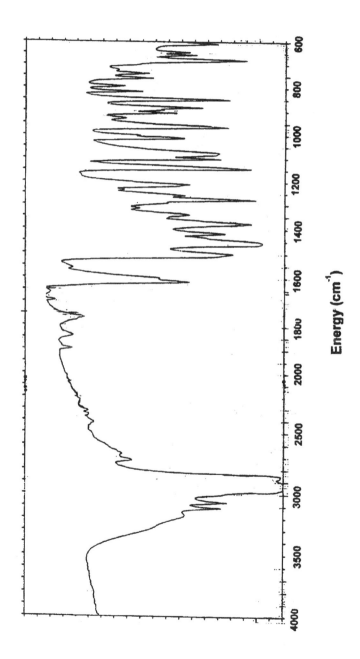

Table 4

Assignment for the Vibrational Transitions of Fluconazole

Energy (cm^{-1})	Assignments
3200	Broad band due to hydrogen bonded O-H stretching vibrations
3120, 3070	Aromatic C-H stretching vibrations
1900, 1845, 1770	Overtone and combination bands consistent with 1,2,4-trisubstitution of phenyl group
1620, 1600, 1510	Aromatic C=C and C=N stretching vibrations
1220, 1210	Aromatic C-F stretching vibrations
1150	C-O stretching vibration consistent with a tertiary alcohol
850	Out-of-plane C-H deformation vibration for two adjacent aromatic hydrogen

Figure 14.   ^{1}H-NMR spectrum of fluconazole in DMSO-D$_6$ at 250 MHz.

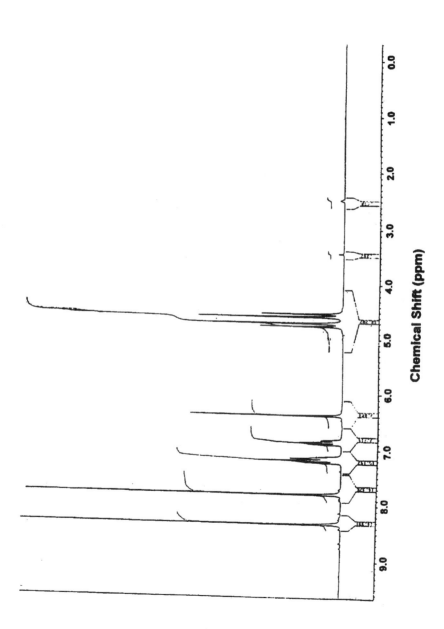

Figure 15. ^{1}H-NMR spectrum of fluconazole in DMSO-D$_6$ containing D$_2$O at 250 MHz (deuterium exchange spectrum).

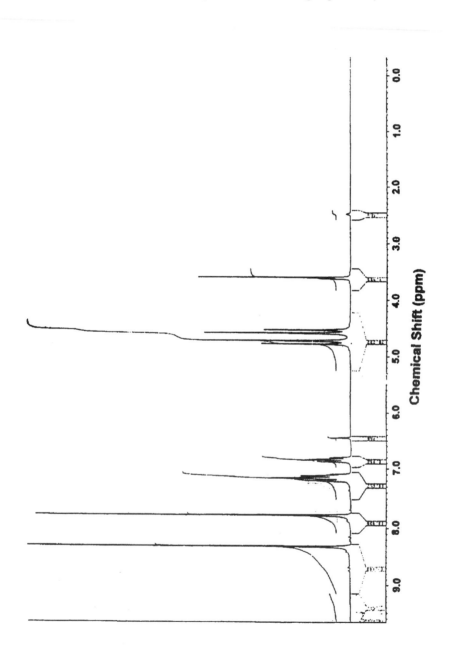

Table 5

Assignment for the ^{1}H-NMR Resonance Bands of Fluconazole

Chemical Shift (ppm)*	Multiplicity	Number of Protons	Assignment	Effect of Deuterium Oxide Addition
2.47	quintet	-	DMSO- d$_5$ (solvent impurity)	--
3.44	singlet	-	HOD	Shifts downfield; More intense
4.50 - 4.80	quartet	4	d, e	No observable change
6.38	singlet	1	j	Substantially reduced intensity
6.80 - 6.93	complex	1	a	No observable change
7.10 - 7.26	complex	2	b, c	No observable change
7.80	Singlet [#]	2	f, i	No observable change
8.34	Singlet [#]	2	g, h	No observable change

* The chemical shifts are reported relative to tetramethylsilane.

Assigned by Nuclear Overhauser Effect Difference Spectroscopy.

3.6.3.2 ^{13}C-NMR Spectrum

The proton decoupled carbon-13 NMR spectrum of fluconazole was obtained using a Bruker WM-250 NMR spectrometer [8]. The spectrum recorded for a 50 mg/mL solution in deuterated dimethyl sulfoxide (DMSO-D$_6$) is shown in Figure 16. The chemical shifts, multiplicities, and assignments, are summarized in Table 6, and make use of the following numbering scheme:

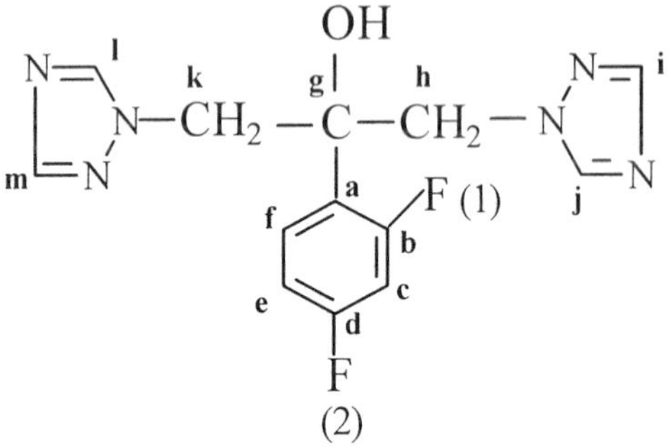

3.6.3.3 ^{19}F-NMR Spectrum

The ^{19}F-NMR spectrum of fluconazole was obtained using Bruker AM 400 and AC 200 cryospectrometers. The fluorine (376.5 MHz) spectra were acquired with a 5-mm probe, using appropriate bandpass and bandstop filters, with deuterated dimethyl sulfoxide (DMSO-D$_6$) being used as the solvent. Based upon the numbering scheme:

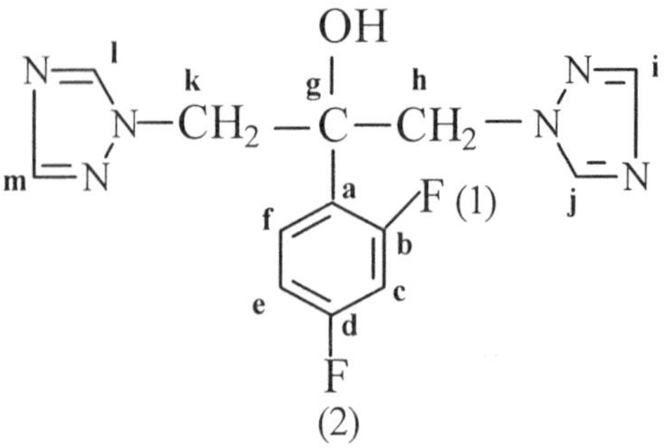

The chemical shift of F(1) was reported to be –106.93 ppm (with a coupling constant of 8.1 Hz) and for F(2) was –110.93 ppm. These chemical shifts are reported relative to an external CFCl$_3$ reference [10].

Figure 16. ^{13}C-NMR spectrum of fluconazole in DMSO-D$_6$ at 62.9 MHz.

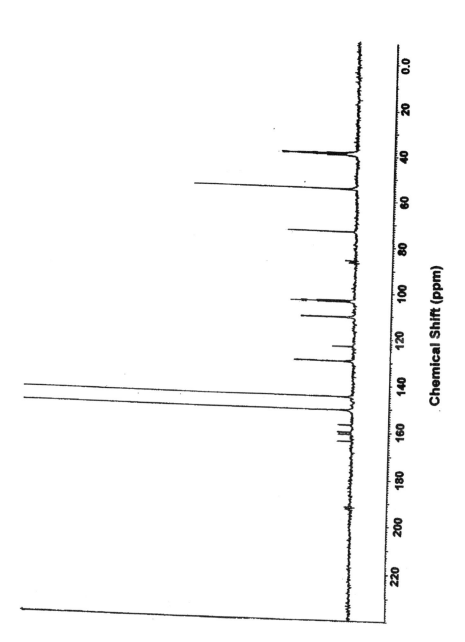

Table 6

Assignment for the ^{13}C-NMR Resonance Bands of Fluconazole

Chemical Shift (ppm)	Multiplicity	Carbon-Fluorine Coupling Constants (Hz)	Assignments
162.0	Doublet of doubles	$C_d - F_2 = 246.6$ $C_d - F_1 = 12.6$	d
159.0	Doublet of doublets	$C_b - F_1 = 247.0$ $C_b - F_2 = 12.2$	b
150.7	singlet	−	j, m
145.1	singlet	−	i, l
129.6	Doublet of doublets	$C_f - F_1 = 9.2$ $C_f - F_2 = 6.1$	f
123.3	Doublet of doublets	$C_a - F_1 = 13.1$ $C_a - F_2 = 3.4$	a
110.8	Doublet of doublets	$C_e - F_2 = 21.1$ $C_e - F_1 = 1.9$	e
103.9	Pseudo-triplet	$C_c - F_1 \atop C_c - F_2 \Big\} = 26.9$	c
73.6	doublet	$C_g - F_1 = 4.9$	g
54.8	doublet	$C_h - F_1 \atop C_k - F_1 \Big\} = 4.9$	h, k
39.4	septet	−	DMSO-D$_6$ (solvent)

3.6.4 Mass Spectrometry

Mass spectra of fluconazole were obtained using a Finnigan Mat 4610 mass spectrometer, operated in the EI mode and employing a gas chromatograph (column type DB-5, J&W Scientific, 15 m x 0.25 i.d., 0.25 μm film thickness). The sample was dissolved in methanol (1 mg/mL). The injection temperature in the GC was maintained at 250°C. The oven temperature program used heating at 200°C (1 minute), then increased at 20°C/min to 280°C (in a period of 15 minutes). The mass spectrum of fluconazole obtained under these conditions is shown in Figure 17. The mass spectrum obtained in this mode did not show a molecular ion (M$^+$) at m/z = 306, and the highest mass fragment was observed at 224 Da [10]. This ion arises by facile cleavage of either triazole group.

The mass spectrum of fluconazole reported in the Wiley Registry of Mass Spectral Data [11] reports additional peaks at m/z of 142 and 170, which are not present in GC/MS spectra reported by Cyr *et al.* [10]. Brammer and co-workers have reported a pseudo-molecular ion (M^{+1} = 307 Da) for fluconazole during LC-thermospray MS analysis [12].

4. Methods of Analysis

4.1 Compendial Tests

Fluconazole is not a compendial article in either the USP or EP, and consequently there is no body of official analytical tests as of yet.

4.2 Elemental Analysis

The theoretical elemental composition of fluconazole based on the molecular formula $C_{13}H_{12}F_2N_6O$ is: C 50.98%, H 3.95%, F 12.41%, N 27.44%, O 5.22% [1].

4.3 Spectrophotometric Methods of Analysis

Spectrophotometric methods have been utilized for the determination of minimum inhibitory concentration (MIC) of fluconazole and other antifungal agents [13-17].

Figure 17.   Direct exposure probe mass spectrum of fluconazole in the
                EI mode.

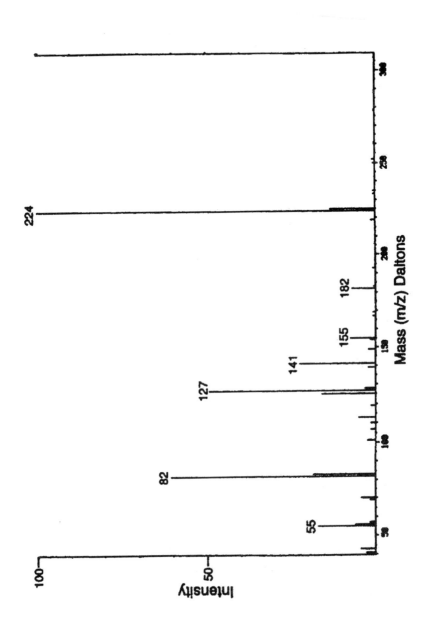

4.4   Gas Chromatography

Gas chromatography (GC) has been utilized for the analysis of fluconazole in biological fluids [18-20]. The various conditions used in those GC methods are summarized in the Table 7.

4.5   High Performance Liquid Chromatography

Several LC methods have been developed for the analysis of fluconazole in pharmaceutical formulations [21,22]. HPLC methods have also been reported for the analysis of fluconazole in biological samples [23-26]. The chromatographic conditions used in these methods are summarized in Table 8.

4.6   Determination in Body Fluids and Tissues

Sane *et al.* have used a bioassay method for the analysis of fluconazole in pharmaceutical preparations [27]. A bioassay method was also used to quantitate fluconazole in animal studies, including rats [28] and mice [29]. Bioassay methodology was used to determine fluconazole concentrations in saliva obtained from AIDS patients [30]. Rex and coworkers also developed a fluconazole bioassay for the analysis of the drug in the plasma, serum, and cerebral spinal fluid, and compared these results with those obtained by HPLC [24].

5.   <u>Stability</u>

5.1   Solution-Phase Stability

Commercially available parenteral formulations of fluconazole have been reported to be stable for 1-2 years, when stored in glass or plastic bottles at 5-30°C [4]. Fluconazole has been found to be very stable in a oral liquid formulation (1 mg/mL) when stored at 4, 23, and 45°C over a period of 15 days [31]. Short term stability studies of fluconazole in injectable solutions have shown these to be stable [32].

Table 7

Summary of the GC Methods and Conditions for the Analysis
of Fluconazole

Conditions	Flow Rate (mL/min)	Column	Detector	Application [reference]
Inject. Temp: 280°C Oven Temp: 230°C Detector Temp: 285°C	$N_2 = 40$ $H_2 = 25$ Air $= 300$	Fused silica DB-5 capillary (15m x 0.32 mm), coated with 0.25 μm 5% methylphenyl: 95% dimethyl-polysiloxane	Nitrogen selective	Serum [18]
Inject. Temp: 225°C Column. Temp: 190°C Detector Temp: 275°C	Not Available	Fused silica DB-5 Megabore. (45 m x 0.55 mm) coated with 1.0 μm bonded liquid phase	Nitrogen phosphorus	Serum Urine Plasma [19]
Inject. Temp: 310°C Oven Temp: initial 100°C for 1 min, and raised to 270°C at a rate of 50°C/min Detector Temp: 350°C	5% methane in Ar carrier gas $= 3$; Aux. Gas $= 60$; Air $= 300$	Fused silica capillary (15m x 0.32 mm) packed with DB-17 (50% phenyl-50% methyl-silicone) and film thickness of 0.5 μm.	Electron capture	Plasma [20]

Table 8

Summary of the HPLC Methods and Conditions for the
Analysis of Fluconazole

Mobile Phase	Flow Rate (mL/min)	Column	Detector	Application [reference]
Acetonitrile: water (15:85 v/v)	1.5	C-18 (15 x 0.6 cm id)	UV at 210 nm	Serum [23]
Methanol: 0.025M Phosphate Buffer (45:55 v/v) pH =7.0	1.0	C-18 (25 x 0.4 cm id)	UV at 260 nm	Plasma, Serum CSF [24]
Acetonitrile: 0.051 M KH_2PO_4 Buffer (15:85 v/v) pH 3.0	0.9	Mixed-phase PTHAA-5 (15 x 0.4 cm id)	UV at 210 nm	Plasma, Serum CSF [25]
Acetonitrile: 0.05 M Na_2HPO_4 Buffer (12:88 v/v) pH 4.0	Not Available	C-8 (15 x 0.39 cm id)	UV at 210 nm	Plasma [26]

5.2   Stability in Formulations

The chemical stability and physical compatibility of fluconazole in an adult total parenteral nutrient solution were investigated under refrigerated conditions over a period of 17 days. A decrease in less than 3% of the fluconazole concentration was noticed in all the admixtures studied [33].

5.3   Incompatibilities with Functional Groups

Lor *et al.* have reported the formation of an immediate precipitate when fluconazole was mixed with ceftazidime, ceftriaxone, cefuroxime, clindamycin, diazepam, and erythromycin. Delayed precipitation, color change, or both, were noticed when this drug was mixed with amphotericin B, cefotaxime, furosemide, or haloperidol. Fluconazole solutions liberate gases when mixed with chloramphenicol or digoxin. A gel-like product was also obtained when this drug was mixed with piperacillin or ticarcillin [34].

6.   Pharmacokinetics and Metabolism

6.1   Animal Studies

A comparison of the various pharmacokinetic parameters of fluconazole in six different animal species have been reported by Jezequel [35], and are summarized in Table-9.

The penetration of fluconazole into the cerebral spinal fluid of rabbits was investigated by Perfect and Durack [36]. The degree of penetration was calculated as:

$$\{ \ [Drug_{CSF}] \ / \ [Drug_{SERUM}] \ \} \ * \ 100$$

This quantity was found to be very high (58-66%).

Table 9

Summary of Mean Pharmacokinetic Parameters of Fluconazole
Obtained in Several Animal Species

Species	Elimination half life (hours)	Distribution Volume, V_d, (L/kg)	Total Body Clearance, (mL/min)	Renal Clearance, (mL/min)
Mouse	4.8	1.1	0.08	0.06
Rat	4.0	0.80	0.22	0.18
Guinea-pig	5.2	0.75	0.64	0.29
Rabbit	10.4 ± 2.3	0.88 ± 0.04	3.0 ± 0.4	No Data
Cat	11.0 ± 1.5	0.50 ± 0.08	1.5 ± 0.4	No Data
Dog	14.0	0.70	8.4	6.0

6.2   Normal Volunteer Studies

Ripa and coworkers have studied the pharmacokinetics of fluconazole in normal volunteers [37]. Pharmacokinetic profiles were found to be independent of the route of administration. This drug was well absorbed orally and excreted unchanged in high concentration in the urine. Fluconazole was found to be metabolically stable. It has an extensive distribution in the extravascular tissue, and has a long plasma elimination half-life.

When 100 mg of the drug was administered by the iv route, the half life was found to be 29.73 ± 8.05 hours. The post-distributive volume was calculated to be 52.16 ± 9.83 L. Around $61.64 \pm 8.80\%$ of the unchanged drug was excreted in the urine after 48 hours. The renal clearance and the plasma clearances were 12.91 ± 2.83 mL/min and 21.03 ± 5.07 mL/min, respectively. When the drug was administered orally as a 50 mg and 150 mg dosages, the distribution and elimination were similar to the iv dose. The peak levels in plasma at 2.5 hours were 0.93 ± 0.13 µg/mL and 2.69 ± 0.43 µg/mL, respectively. Lazar and Hilligoss [38] have reported that the oral bioavailability of fluconazole is 90%, and that plasma protein binding is only 12%. The volume of distribution approximates to that of total body mass. Fluconazole is metabolically very stable, and only 11% of a single dose is excreted as metabolites in the urine.

The mean cerebral spinal fluid penetration of fluconazole (defined by the ratio of the concentration in cerebral spinal fluid to the concentration of serum) after 50 mg iv administration was 0.52. For a dose of 100 mg iv, this ratio was reported to be 0.62. The penetration after oral administration was also reported to be high (greater than 0.52 to 0.93).

6.3   Patient Studies

The plasma clearance of fluconazole may be lowered in patients with HIV infection when compared to immuno-competent patients. The elimination half-life may also be prolonged in such patients [39,40]. The pharmacokinetics of fluconazole is markedly affected by impaired renal function [41,42]. The pharmacokinetics of this drug in the cerebral spinal fluid and serum of humans with coccidioidal meningitis has been reported by Tucker et al. [43]. The penetration of the drug into the cerebral spinal fluid was reported to be substantially high (74-89%) in those patients.

7. Toxicity

Very limited information is available on the acute toxicity of fluconazole in humans. In rats and mice, a high dose of the drug may produce decreased motility and respiration [4]. No fatalities in those animals were noticed when 1 g/kg of fluconazole was administered. However, at a high dose (1-2 g/kg), death occurred 1.5 hours to 3 days after the dose (preceded by clonic seizures). The oral LD_{50} (mice/rat) is 1270-1520 mg/kg [44].

No evidence of mutagenesis was noticed when fluconazole was tested with *Salmonella typhimurium*, or in the mouse lymphoma L5178Y system [4]. No evidence of carcinogenicity was found in rats after they received 2-7 times the human dose of the drug by oral route, over a period of 24 months. However, incidence of hepatocellular adenoma was increased in male rats receiving an oral dose of 5 or 10 mg/kg/day was reported [4]. Fluconazole has also been reported to be less immunotoxic than itraconazole [45].

8. Drug Interactions

Drug-drug interaction between fluconazole and some other commonly used drug have been reported. Some of the drug-drug interactions may affect clinical outcome, and must therefore be carefully evaluated during the management of antifungal therapy. Interaction of fluconazole with rifampin, phenytoin, carbamazepine, cyclosporin, oral contraceptive, antacid, coumadin, and sulfonyl urea have been reported else where [46,47].

Acknowledgement

The authors would like to thank Genevieve Mylott (Pfizer Inc, New York) for providing some technical data on fluconazole, which has been referred herein as 'unpublished data from Pfizer Inc.'

References

1.   *The Merck Index*, S. Budavari, ed., 11th Edn., Merck and Co., Inc., Rahway, N.J., USA, 1989, p. 4058.

2.   "Drug Discovery: The Discovery of Fluconazole", *Pharmaceutical News*, **2**, 9-12 (1995).

3.   J. E. Bennett, "Antimicrobial Agents: Antifungal Agents", in *Goodman And Gilman's Pharmacological Basis of Therapeutics*, 8th Edition, A. G. Gilman, ed., Pergamon Press, New York, 1990, p. 1170.

4.   *AHFS Drug Information*, G.K. McEvoy, ed., The American Society of Health-System Pharmacists, Maryland, 1998, pp. 96-106.

5.   K. Richardson, K. Cooper, M. S. Marriott, M.H. Tarbit, P. F. Troke, and P. J. Whittle, *Rev. Infect. Dis.*, **12**, S267-271 (1990).

6.   K. Richardson, K. Cooper, M. S. Marriott, M. H. Tarbit, P. F. Troke, and P. J. Whittle, *Ann. N. Y. Acad. Sci.*, **544**, 4-11 (1988).

7.   A. K. Dash, unpublished data.

8.   Unpublished Data from Pfizer, Inc.

9.   D. Stopher and S. McClean, *J. Pharm. Pharmacol.*, **42**, 144 (1990).

10. T. D. Cyr, B. A. Dawson, G. A. Neville, and H. F. Shurvell, *J. Pharm. Biomed. Anal.*, **14**, 247-255 (1996).

11. *Wiley Registry of Mass Spectral Data*, 6th Edition, Wiley, New York, 1994.

12. K. W. Brammer, J. A. Coakley, S. G. Jezequel, and M. H. Tarbit, *Drug. Metab. Dispos.*, **19**, 764-767 (1991).

13. J. L. Rodriguez-Tudela, and J. V. Martinez-Suarez, *J. Antimicrob. Chemotherapy*, **35**, 739-749 (1995).

14.   M.L. Pfaller, S. A. Messer and S. Coffmann, *J. Clin. Microbiol.*, **33**, 1094-1097 (1995).

15.   A. Espinenel-Ingroff, J. L. Rodriguez-Tudela, and J. V. Martinez-Suarez, *J. Clin. Microbiol.*, **33**, 3154-3158 (1995).

16.   F. C. Odds, L. Vranckx, and F. Woestenborghs, *Antimicrob. Agents. Chemotherapy*, **39**, 2051-2060 (1995).

17.   M. H. Nguyen and C. Y. Yu, *J. Clin. Microbiol.*, **37**, 141-145 (1999).

18.   D. Debruyne, J. P. Ryckelynck, M. C. Bigot, and M. Moulin, *J. Pharm. Sci.*, **77**, 534-535 (1988).

19.   S. C. Harris, J. E. Wallace, G. Foulds and M. G. Rinaldi, *Antimicrob. Agents Chemotherapy.*, **33**, 714-716 (1989).

20.   A. B. Rege, J. Y. Walker-Cador, R. A. Clark, J. J. L. Lertora, N. E. Hyslop Jr., and W. J. George, *Antimicrob. Agents Chemotherapy*, **36**, 647-650 (1992).

21.   W. P. Xu, B. C. Hu, P. Cheng, and L. M. Chen, *Chin. J. Pharm. Anal.*, **17**, 313-315 (1997).

22.   R.A. Ashworth, L. J. Pescko, H. T. Karnes, and D. Lowe, *ASHP Midyear Clinical Meeting*, **26**, P-572E (1991).

23.   K. K. Hosotsubo, H. Hosotsubo, M. K. Nishijima, T. Okada, N. Taenaka, and I. Yoshiya, *Br. J. Chrom.*, **529**, 223-228 (1990).

24.   J. H. Rex, L. H. Hanson, M. A. Amantea, D.A. Stevenes, and J. E. Bennett, *Antimicrob. Agents Chemotherapy*, **35**, 846-850 (1991).

25.   J. E. Wallace, S. C. Harris, J. Gallegos, G. Foulds, T. J. H. Chen, and M. G. Rinaldi, *Antimicrob. Agents Chemotherapy*, **36**, 603-606 (1992).

26.   F. J. Flores-Murrieta, V. Granados-Soto and E. Hong, *J. Liquid Chrom.*, **17**, 3803-3811 (1994).

27.   R. T. Sane, A. A. Fulay, and A. N. Joshi, *Indian Drugs,* **31**, 207-210 (1994).

28.   J. Van-Cutsem, *Mycoses,* **32**, 14-34 (1989).

29.   J. S. Hostetler, L. H. Hanson and D. A. Stevens , *Antimicrob. Agents Chemotherapy,* **36**, 477-480 (1992).

30.   D. Gracia-Hermoso, F. Dromer, L. Improvisi, F. Provost, and B. Dupont. *Antimicrob. Agents Chemotherapy,* **39**, 656-660 (1991).

31.   W. Yamreudeewong, A. Lopez-Anaya, and H. Rappaport, *Am. J. Hosp. Pharm.,* **50**, 2366-2367 (1993).

32.   A. K. Hunt-Fugate, C. K. Hennessey, and C.M. Kazarian, *Am. J. Hosp. Pharm.,* **50**, 1186-1187 (1993).

33.   W. R. Outman, R. J. Baptista, F. P. Mitrano, and D. A. Williams, *ASHP Annual Meeting,* **48**, P-63E (1991).

34.   E. Lor, T. Sheybani and J. Takagi, *Am. J. Hosp. Pharm.,* **48**, 744-746 (1991).

35.   S. G. Jezequel, *J. Pharm. Pharmacol.,* **46**, 196-199 (1994).

36.   J. R. Perfect and D. T. Durack, *J. Antimicrob. Chemotherapy,* **16**, 81-86 (1985).

37.   S. Ripa, L. Ferrante and M. Prenna, *Chemotherapy,* **39**, 6-12 (1993).

38.   J. D. Lazar and D. M. Hilligoss, *Seminars in Oncology,* **17**, 14-18 (1990).

39.   S. Tett, S. Moore and J. Ray, *Antimicrob. Agents Chemotherapy,* **39**, 1835-1841 (1995).

40.   A. J. McLachlan and S. Tett, *Br. J. Clin. Pharmacy,* **41**, 291-298 (1996).

41.   S. Toon, C. E. Ross, R. Gokal, and M. Rowland, *Br. J. Clin. Pharmacy,* **29**, 221-226 (1990).

42. S. Oono, K. Tabei, T. Tetsuka, and Y. Asano, *Eur. J. Clin. Pharmacol.*, **42**, 667-669 (1992).

43. R. M. Tucker, P. L. Williams, E. G. Arathoon, B. E. Levine, A. I. Hartstein, L. H. Hanson, and D. A. Stevens, *Antimicrob. Agents Chemotherapy*, **32**, 369-373 (1988).

44. ***Fluconazole: Material Safety Data Sheet***, Pfizer Inc.; http://msds.pdc.cornell.edu/msds/siri/q259/q351.html.

45. V. Vuddhakul, G. T. Mai, J. G. McCormack, W. K. Seow, and Y. H. Thong, *Int. J. Immunopharmacy*, **12**, 639-645 (1990).

46. R. M. Tucker, D. W. Denning, L. H. Hanson, M. G. Rinaldi, J. R. Graybill, P. K. Sharkey, D. Pappagianis, and D. A. Stevens, *Clin. Infect. Dis.*, **14**, 165-174 (1992).

47. S. F. Kowalsky, *Pharmacotherapy*, **10**, 170S-173S (1990).

FLUTAMIDE

Richard Sternal and Niran Nugara

Analytical Development

Schering-Plough Research Institute

Kenilworth, NJ   07033

ANALYTICAL PROFILES OF
DRUG SUBSTANCES AND EXCIPIENTS
VOLUME 27

Contents

1.   Description

1.1   Nomenclature

1.1.1   Chemical Names

2-Methyl-*N*-[4-nitro-3-(trifluoromethyl)phenyl]propanamide

α,α,α-trifluoro-2-methyl-4'-nitro-m-propionotoluidide

4'-nitro-3'-trifluoromethylisobutyranilide

1.1.2   Nonproprietary Names

Flutamide

1.1.3   Proprietary Names

Drogenil, Eulexin, Euflex, Flucinom, Flugeril, Fugerel, Sebatrol

1.2   Formulae

1.2.1   Empirical

$C_{11}H_{11}F_3N_2O_3$

1.2.2   Structural

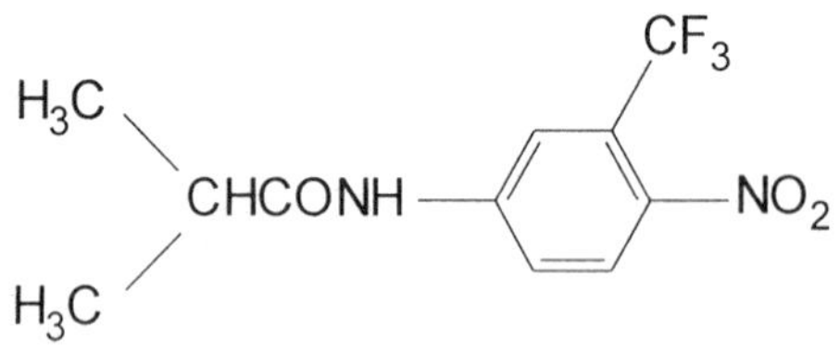

1.3   Molecular Weight

276.22

1.4   CAS Number

13311-84-7

1.5   Appearance

Flutamide is a pale yellow crystalline powder [1].

1.6   Uses and Applications [2, 3, 4]

Flutamide is an acetanilide, nonsteroidal, orally active anti-androgen. It exerts its anti-androgenic action by inhibiting androgen uptake and/or by inhibiting nuclear binding of androgen in target tissues. Prostatic carcinoma is known to be androgen-sensitive and responds to treatment that counteracts the effect of androgen and/or removes the source of androgen (*e.g.*, castration). It is indicated for use in combination with LHRH agonists for the management of locally confined Stage B_2-C and Stage D_2 metastatic carcinoma of the prostate.

2.   Methods of Preparation

Three general methods have been described for the synthesis of flutamide.

Flutamide may be prepared according to the procedure outlined in Scheme 1 [5]. Isobutyryl chloride is added in a slow dropwise fashion to a stirred cooled solution of 4-nitro-3-trifluoromethyl-aniline in pyridine. The reaction mixture is then heated in a steam bath. The resulting mixture is cooled and poured into ice water and filtered. The crude anilide is washed with water and crystallized from benzene to obtain analytically pure material.

Scheme 1. The synthesis of flutamide by condensation.

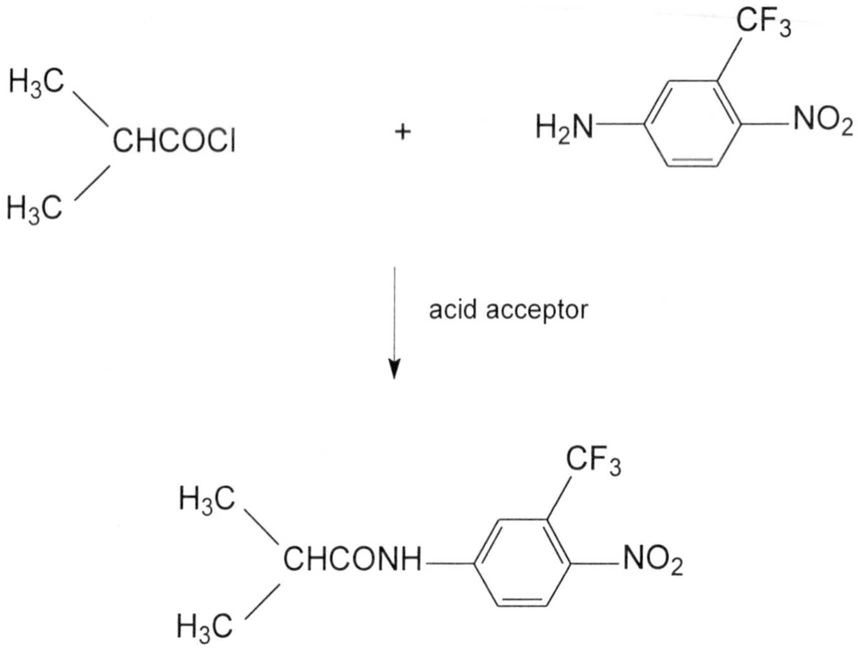

acid acceptor

An alternate method of synthesis is outlined in Scheme 2 [6]. Slowly add *m*-trifluoromethylisobutyranilide to 15-18% oleum maintained at 5°C. Add dropwise 90% nitric acid to this stirred mixture kept at 5°C. The mixture is then poured into ice water with stirring and filtered. The product is washed with water to remove excess acid and dried.

Flutamide and a series of closely related analogs were synthesized using the procedure outlined in Scheme 3 [7]. This process uses the reaction of thionyl chloride in dimethylacetamide at -20°C to generate the acid chloride *in situ*. This is coupled with the appropriate aniline to yield the desired anilide.

Scheme 2. The synthesis of flutamide by para-nitration.

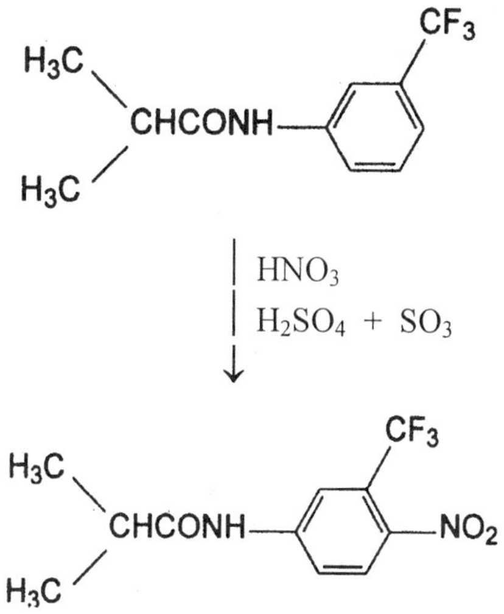
CF3
H3C
CHCONH
H3C
HNO3
H2SO4 + SO3
CF3
H3C
CHCONH
NO2
H3C

Scheme 3. The synthesis of flutamide by *in situ* generation of acid chloride.

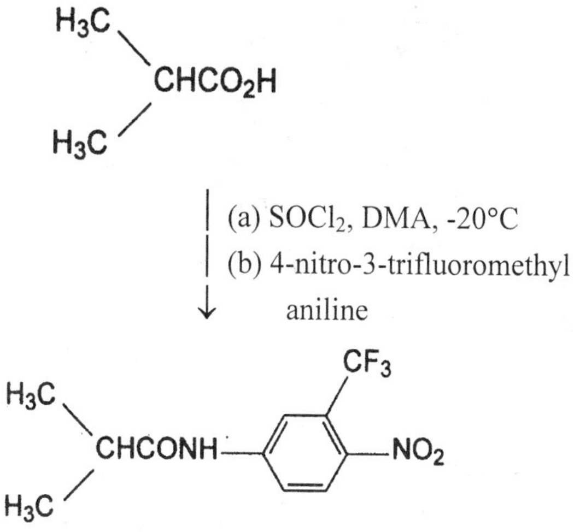
H3C
CHCO2H
H3C
(a) SOCl2, DMA, -20°C
(b) 4-nitro-3-trifluoromethyl
aniline
CF3
H3C
CHCONH
NO2
H3C

3.   Physical Properties

3.1   Particle Morphology

A photomicrograph was obtained using a Nikon Eclipse E800 system at 100X, with the sample dispersed in Dow 200 oil. As the photomicrograph in Figure 1 indicates, flutamide particles have an elongated (needlelike) shape, and the individual particle size is in the 25 μm range.

The particle size distribution was determined using a Sympatec HELOS Laser Diffraction Particle Size Analyzer, with the sample being suspended in 0.1% polysorbate 80. The results are summarized in Figure 2, and yield a mean particle size of 22.5 μm.

3.2   Crystallographic Properties

3.2.1   Single Crystal Structure

The single crystal structure determination of flutamide has been reported [8]. Single crystals were grown by evaporating a saturated solution of flutamide in ethanol. The x-ray diffraction intensities of the single crystal planes were collected on a CAD-4 ENRAF-NONIUS diffractometer. The structure was elucidated using direct methods with the program MULTAN 11/82, using the x-ray diffraction experimental parameters in Table 1. The crystal structure thus obtained is shown in Figure 3, and the stereoview of the crystal packing is shown in Figure 4.

3.2.2   Polymorphism

A single polymorph of flutamide is consistently produced. It was reported [9] that the amorphous state (T_g = 272 K) may be obtained by quenching the melt. DSC studies indicate the existence of another polymorph (metastable) form, which is transformed into the stable form at room temperature, but this metastable polymorph was not isolated.

3.2.3   X-Ray Powder Diffraction Pattern

The x-ray powder diffraction pattern of flutamide was obtained using a Philips PW1710 x-ray powder diffraction system. The radiation source used was the K-α line of copper at 1.54056 Å, operated at 45 kV and 40

Figure 1.   Photomicrograph of flutamide obtained at a magnification
             of 100X.

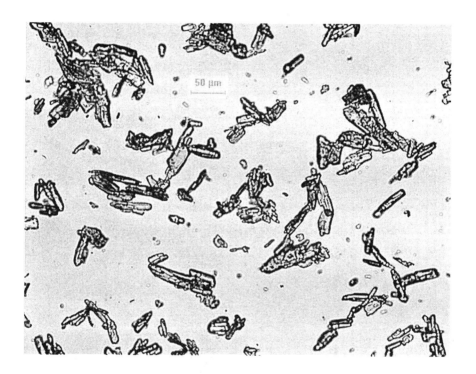

Figure 2.   Particle size distribution for flutamide.

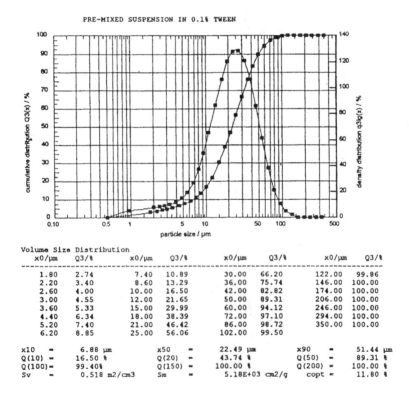

PRE-MIXED SUSPENSION IN 0.1% TWEEN

Volume Size Distribution

x0/μm	Q3/%	x0/μm	Q3/%	x0/μm	Q3/%	x0/μm	Q3/%
1.80	2.74	7.40	10.89	30.00	66.20	122.00	99.86
2.20	3.40	8.60	13.29	36.00	75.74	146.00	100.00
2.60	4.00	10.00	16.50	42.00	82.82	174.00	100.00
3.00	4.55	12.00	21.65	50.00	89.31	206.00	100.00
3.60	5.33	15.00	29.99	60.00	94.12	246.00	100.00
4.40	6.34	18.00	38.39	72.00	97.10	294.00	100.00
5.20	7.40	21.00	46.42	86.00	98.72	350.00	100.00
6.20	8.85	25.00	56.06	102.00	99.50		

x10	=	6.88 μm	x50	=	22.49 μm	x90	=	51.44 μm
Q(10)	=	16.50 %	Q(20)	=	43.74 %	Q(50)	=	89.31 %
Q(100)	=	99.40%	Q(150)	=	100.00 %	Q(200)	=	100.00 %
Sv	=	0.518 m2/cm3	Sm	=	5.18E+03 cm2/g	copt	=	11.80 %

Table 1

Single Crystal Data for Flutamide [8]

Parameter	
Molecular formula	$C_{11}H_{11}F_3N_2O_3$
Space group	$Pna2_1$
F(000)	568
Z	4
μ	1.21 mm^{-1}
a(Å)	11.856(2)
B	20.477(3)
C	4.9590(9)
V(Å^3)	1203.9(6)
Dcal (Mg m^{-3})	1.52
Crystal size (mm)	0.45 x 0.06 x 0.05
Morphology	Yellow needles
Radiation, Å	Cu Kα, λ = 1.5418
Temperature (°K)	294
Observed reflections	859 ($I \geq 2\sigma(I)$)
R factor	0.034
R factor (weighed)	0.040
W	$1/\sigma^2(F)$

Used with permission from Plenum Publishing.

Figure 3.   Structure of flutamide elucidated from single crystal x-ray data, obtained from reference [8] and used with permission from Plenum Publishing.

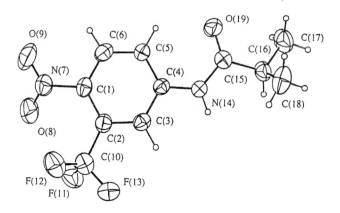

Figure 4.   Stereoview of flutamide molecules in the unit cell. Dotted lines represent hydrogen bonds. Obtained from reference [8] and used with permission from Plenum Publishing.

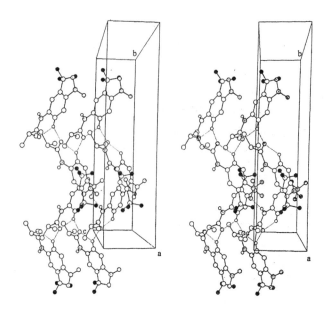

Figure 5.   X-ray powder diffraction pattern of flutamide.

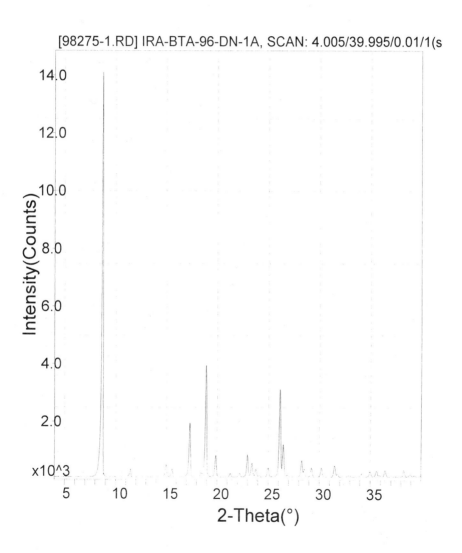

Table 2

Data from the X-Ray Powder Diffraction Pattern of Flutamide

Scattering Angle (degrees 2-θ)	d-spacing (Å)	Relative Intensity, I/I$_0$ (%)
4.515	19.5539	0.1
8.705	10.1496	100.0
8.856	9.9769	0.7
11.465	7.7118	2.1
14.986	5.9070	3.2
15.585	5.6811	2.0
17.325	5.1143	13.0
18.454	4.8039	1.0
18.915	4.6878	27.2
19.875	4.4636	5.3
21.324	4.1633	0.9
22.185	4.0037	1.0
22.985	3.8662	5.4
23.435	3.7929	3.5
23.815	3.7332	1.7
24.136	3.6843	0.2
25.004	3.5582	2.3
26.145	3.4056	22.0
26.465	3.3651	7.9
26.863	3.3161	0.2
27.176	3.2786	0.1

Table 2 (continued)

Data from the X-Ray Powder Diffraction Pattern of Flutamide

28.295	3.1515	4.1
28.554	3.1234	2.2
29.245	3.0512	2.3
30.185	2.9583	2.3
30.465	2.9317	0.3
31.475	2.8399	2.9
31.766	2.8146	0.9
32.687	2.7373	0.2
32.925	2.7181	0.6
33.847	2.6462	0.3
34.184	2.6208	0.9
34.976	2.5633	1.5
35.645	2.5167	1.5
35.933	2.4971	0.4
36.405	2.4658	1.6
37.355	2.4053	0.2
38.285	2.3490	1.8
38.935	2.3113	0.6
39.314	2.2898	0.3

mA. The sample was scanned between 4.005 and 39.995 degrees 2-θ, in step sizes of 0.01° at a rate of 0.01° 2-θ / second. The powder pattern obtained is presented in Figure 5, and a summary of scattering angles, d-spacings and relative intensities is presented in Table 2.

3.3   Thermal Methods of analysis

3.3.1   Melting Behavior

The melting range of flutamide has been reported as 111.5°C to 112.5°C [10].

3.3.2   Differential Scanning Calorimetry

Differential scanning calorimetry analysis was performed using a TA Instruments model 2920 system. Approximately 3 mg of sample was placed on an open aluminum pan, and the pan was heated under nitrogen (purged at ~30 mL/minute) up to 150°C at a rate of 10°C/minute. The DSC thermogram is shown in Figure 6, and exhibits an endothermic transition with an onset temperature of 111.5°C. This feature is assigned to the fusion transition, which is consistent with the above reported value for melting range.

3.3.3   Thermogravimetric Analysis

A thermogravimetric analysis study of flutamide was performed using a DuPont Instruments model 2950 thermal analysis system. A 14 mg sample was heated a rate of 10°C/minute up to a final temperature of 200°C under nitrogen (purged at ~110 mL/minute). As shown in Figure 7, the TGA thermogram exhibits no weight loss attributable to residual solvents, but does begin to exhibit a significant weight loss starting at approximately 120°C. This is due to the thermal decomposition of the compound, and occurs after the onset of DSC melting at 111.5°C.

3.4   Hygroscopicity

Moisture uptake for this drug substance is negligible. Weight changes that are less than 0.5% have been recorded after exposure over the relative humidity range of 0-95%.

Figure 6.   Differential scanning calorimetry thermogram of flutamide.

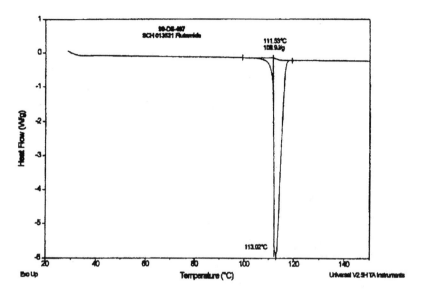

Figure 7.   Thermogravimetric analysis thermogram of flutamide.

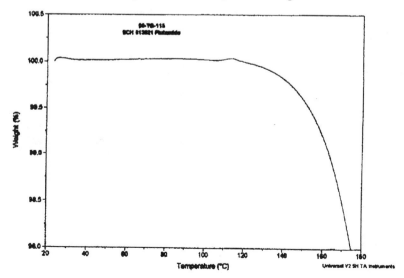

Using a moisture balance (VTI) instrument, equilibrium weight gain was recorded over the range 5% to 95% relative humidity at 25°C for two complete cycles. The result for the first cycle is presented in Figure 8. Both cycles showed a similar continuous and reversible adsorption / desorption behavior with minimal weight gain even at 95% RH. In addition, the adsorption / desorption isotherm shows very little hysteresis, indicating that the flutamide solid state has a low affinity for water. Based on these observations, flutamide is deduced not to be hygroscopic.

Figure 8.   Adsorption/desorption isotherm for flutamide.

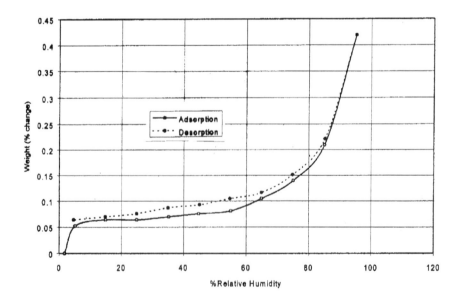

3.5   Solubility Characteristics

Flutamide is insoluble in water, but freely soluble in polar organic solvents such as acetone, alcohol, dimethyl formamide, dimethyl sulfoxide, ethyl acetate, methanol, and polyethylene glycol 400. It is also soluble in chloroform, diethyl ether, and propylene glycol. A summary of the solubility in various solvents is given in Table 3 [17].

3.6   Partition Coefficients

The partition coefficient of flutamide was determined in four different biphasic systems consisting of 1-octanol and an aqueous phase, and the following results were obtained:

Biphasic System	Log P (oil/aqueous)
1-octanol/0.1 N HCl	> 3.4
1-octanol/0.05M phosphate buffer (pH 7)	> 3.4
1-octanol/deionized water	> 3.4
1-octanol/0.1N NaOH	> 2.9

These results show that due to the nonpolar nature and the low solubility of flutamide in aqueous media, it partitions significantly more into the 1-octanol phase than the aqueous phase.

3.7   Ionization Constants

Using the ACD program (Advanced Chemistry Development, Toronto, Canada), ionization constants (pKa) of 13.1 and −6.4 were calculated for the amide group. No experimental determinations in water have been reported.

An apparent pKa in a hydro-alcoholic solution (40 mM Britton-Robinson buffer / ethanol, 80/20) has been determined [11]. The pKa for the amide group was determined by a polarographic method to be 4.75, and 4.83 when determined using UV-VIS spectrophotometry.

Table 3

Solubility of Flutamide in Various Solvents

Solvent System	Solubility (mg/mL, 25°C)	USP Solubility Definition
Acetone	>100	Freely Soluble
Benzene	9.0	Slightly Soluble
Chloroform	65	Soluble
Dimethylformamide	>100	Freely Soluble
Dimethylsulfoxide	>100	Freely Soluble
Ethanol	>100	Freely Soluble
Ethanol 85% / Water 15% (v/v)	>100	Freely Soluble
Ether	70	Soluble
Ethyl Acetate	>100	Freely Soluble
Methanol	>100	Freely Soluble
Mineral Oil	0.04	Insoluble
Petroleum Ether	0.03	Insoluble
Polyethylene Glycol 400	>100	Freely Soluble
Polyethylene Glycol 400, 40% (w/v) – Ethanol 10% (v/v) q.s. Water	1.6	Slightly Soluble
Polyethylene Glycol 400, 50% / Water 50%	1.1	Slightly Soluble
Propylene Glycol	50	Soluble
Propylene Glycol 40% (w/v) / Ethanol 10% (v/v) q.s. Water	0.75	Very Slightly Soluble
Propylene Glycol 50% / Water 50% (w/v)	0.60	Very Slightly Soluble
Water	0.05	Insoluble

3.8   Spectroscopy

3.8.1   UV/VIS Spectroscopy

The ultraviolet spectrum of flutamide in a neutral solution (7:3 v/v methanol / water) was obtained using a Shimadzu spectrophotometer model 1601. The spectrum is shown in Figure 9, and the wavelengths of maximum absorption and the molar absorptivities at each wavelength were determined to be as follows:

Wavelength Maximum (nm)	Molar Absorptivity (L/mole·cm)
296	1.09×10^4
228	7.24×10^3

3.8.2   Vibrational Spectroscopy

The infrared spectrum of a mineral oil mull of flutamide was obtained using a Mattson Model 6021 infrared spectrometer. The spectrum shown in Figure 10 is consistent with the structure of flutamide, and the infrared assignments are presented in Table 4.

3.8.3   Nuclear Magnetic Resonance Spectrometry

3.8.3.1 ^{1}H-NMR Spectrum

A 400 MHz proton NMR of a solution of flutamide at 25°C (dissolved in deuterated DMSO) was obtained using a Varian XL 400 spectrometer. The spectrum is presented in Figure 11, and the chemical shift assignments are given in Table 5.

3.8.3.2 ^{13}C-NMR Spectrum

The ^{13}C-NMR spectrum of flutamide was obtained in deuterated DMSO using a Varian XL 400 spectrometer at 25°C. The full spectrum, with expansion of the 117-126 ppm region, is presented in Figure 12 and the Attached Proton Test (APT) spectrum is presented in Figure 13. The ^{13}C chemical shift assignments (Table 6) were made by using the ^{1}H-NMR spectrum, APT spectrum, and the structure of flutamide.

Figure 9.   UV spectrum of flutamide in 7:3 v/v methanol/water.

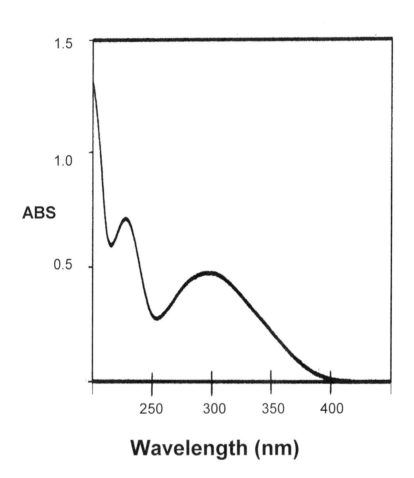

Figure 10. IR spectrum of flutamide (mineral oil dispersion).

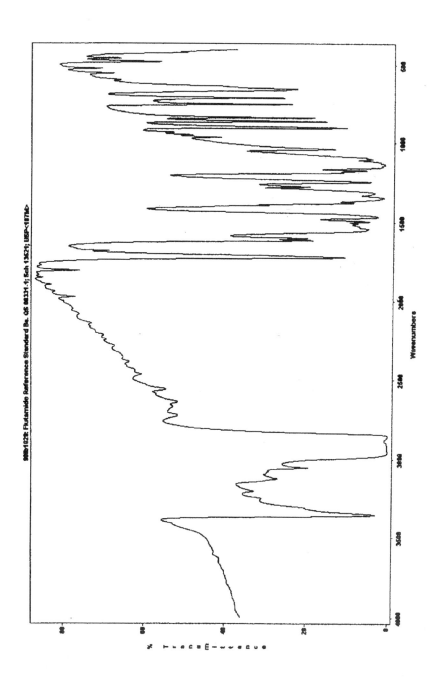

Table 4

Assignments for the Vibrational Transitions of Flutamide

Energy (cm^{-1})	Assignment
3356 (s)	N-H stretch
1717 (m)	C=O stretch
1610, 1597 (m)	C=C stretch
1540 (m)	NO_2 stretch
1344 (s)	CF_3
1315 (m)	NO_2
1243 (m)	C-N (amide)
1136 (s)	CF_3
903, 862 (m)	1,2,4-trisubstituted benzene
754 (m)	C-N (C-NO_2)

Intensities: m = medium
s = strong

Figure 11. ^{1}H-NMR spectrum of flutamide.

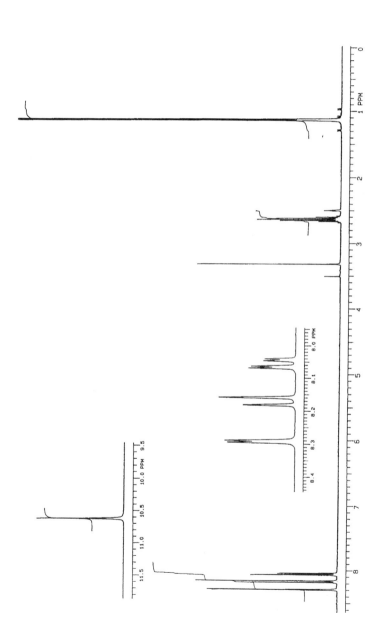

Table 5

^{1}H NMR Chemical Shift Assignments for Flutamide

Protons	Chemical Shifts	Intensity	Multiplicity
CH(C$\underline{H_3}$)$_2$	1.13	6 H	Doublet J = 6.9 Hz
C$\underline{H}$(CH$_3$)$_2$	2.63	1 H	Multiplet
H$_5$	8.17	1 H	Doublet J ortho = 9.0 Hz
H$_2$	8.29	1 H	Broad doublet J meta = 2.2 Hz
H$_6$	8.06	1 H	Doublet of doublet (J = 8.9, 2.2 Hz)
NH	10.62	1 H	Broad singlet

Residual proton resonances from water at 3.28 ppm, DMSO at 2.50 ppm, and a low level impurity at 3.5 ppm are also present.

Figure 12. ^{13}C NMR spectrum of flutamide.

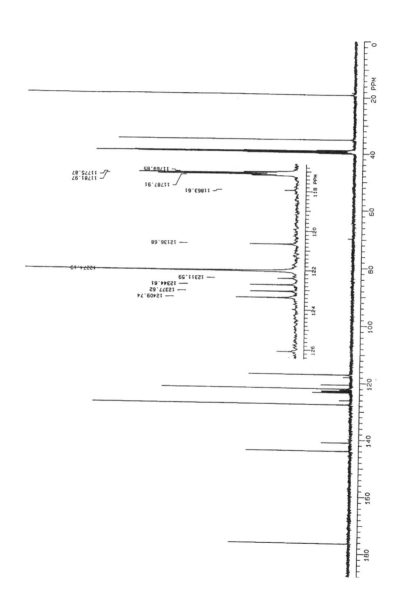

Figure 13.   Attached Proton Test spectrum for flutamide (top).
                  ^{13}C-NMR spectrum for flutamide (bottom).

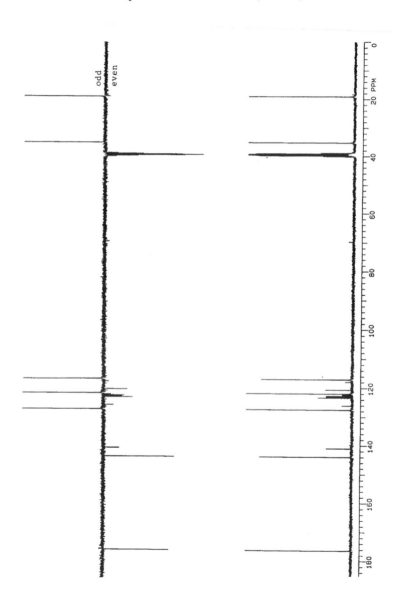

Table 6

^{13}C NMR Chemical Shift Assignments for Flutamide

Carbon	Chemical Shifts (ppm)
CH(C̲H$_3$)$_2$	19.1
C̲H(CH$_3$)$_2$	35.2
C$_2$	117.1
C$_6$	122.0
C$_5$	127.6
C$_3$	126.1
CF$_3$	123.4
C$_4$	141.1
C$_1$	144.0
C = O	176.4

Chemical shifts at 39.5-40 ppm due to DMSO. Low-level impurity chemical shifts are present at 118, 120.7, 122.4, 122.7, and 123 ppm.

3.8.4   Mass Spectrometry

The FAB mass spectrum of flutamide was obtained using a Finnigan MAT 90 instrument. The spectrum is presented in Figure 14, and a listing of fragmentation assignments is given in Table 7.

4.   Methods of Analysis

4.1   Compendial Tests

The compendial tests for flutamide USP are given in the drug substance monograph [12]. The specified testing consists of compound identification, melting range, loss on drying, residue on ignition, heavy metals, chromatographic purity and assay.

4.1.1   Identification

The identification of flutamide can be made on the basis of the equivalence of its infrared spectrum (obtained in nujol) with that of a standard spectrum. Alternatively the identification can be made on the basis of matching the retention time of the major peak in the HPLC chromatogram of the sample with the major peak in the USP Reference Standard using the USP assay procedure.

4.1.2   Melting Range

The melting range is determined according to USP general test <741>, and should be between 110° and 114°C, with a range of not more than 2°C.

4.1.3   Loss on Drying

The loss on drying is determined according to USP general test <731>. The sample is dried at 60°C for 3 hours under vacuum, with a weight loss specification of not more than 1.0%.

4.1.4   Residue on Ignition

The residue on ignition is determined according to USP general test <281>, and should be not more than 0.1%.

Figure 14.   Fast atom bombardment mass spectrum of flutamide.

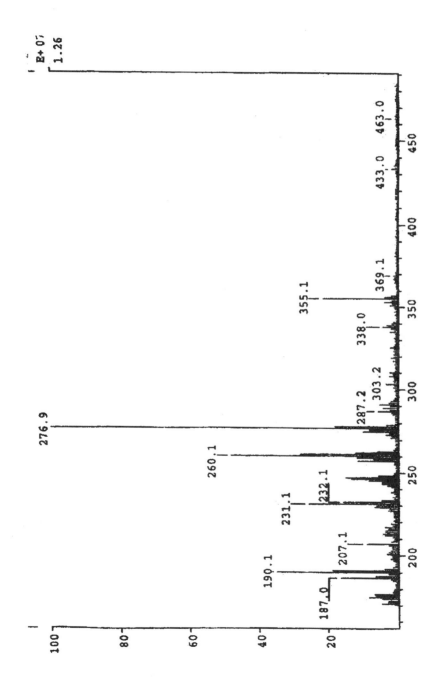

Table 7

Assignments for the FAB Mass Spectrum of Flutamide

Mass (amu)	Ions	Fragment Lost
355	$(MH + DMSO)^+$(a)	
277	MH^+	
261	$(MH - 16)^+$	O
260	$(M - 16)^+$	O
247	$(MH - 30)^+$	C_2H_6
246	$(M - 30)^+$	C_2H_6
232	Background[1] (b)	
231	$(MH - 46)^+$	NO_2
207	Background[1] (b)	
191	$(MH - 86)^+$	C_3H_7CONH
190	$(M - 86)^+$	C_3H_7CONH
188	$(MH - 89)^+$	C_3H_7, NO_2
187	$(M - 89)^+$	C_3H_7, NO_2
175	$(MH - 102)^+$	C_3H_7CONH, O

a: DMSO is the solvent used for the sample preparation.

b: Background fragment ions generated by the instrument.

4.1.5 Heavy Metals

The heavy metals content is determined according to USP general test <231>, Method II, and should be not more than 10 ppm.

4.1.6 Chromatographic Purity

To perform the Chromatographic Purity determination, the following solutions are prepared:

Mobile Phase: Prepare a 1:1 v/v filtered and degassed mixture of deionized water and acetonitrile.

System Suitability Solution: Prepare a solution containing about 0.004 mg/mL of USP o-flutamide reference standard (RS) and 2 mg/mL of USP flutamide RS in mobile phase.

Test Solution: Dissolve 2 mg/mL of flutamide in acetonitrile.

Chromatographic system: Use a 4.6 mm x 25-cm column that contains packing L1 (octadecyl silane chemically bonded to porous silica or ceramic microparticles, 3 to 10 μm in diameter). The mobile phase is eluted at a flow rate of 0.5 mL/minute. and detection is made on the basis of the UV absorbance at 240 nm. The injection volume is 20 μL, and the run time continues until the o-flutamide peaks elutes (RRT = 1.4). The System Suitability Solution must have a resolution of not less than 2.0, and the relative standard deviation for replicate injections of o-flutamide must be not more than 6.0%. The Test Solution is injected and peak responses are recorded. The specification is not more than 0.5% total impurities.

4.1.7 Assay

To perform the Assay determination, the following solutions are prepared:

Mobile Phase: Prepare a 1:1 v/v filtered and degassed mixture of deionized water and acetonitrile.

Standard Preparation: Prepare a 0.5 mg/mL solution of USP flutamide RS in Mobile Phase.

System Suitability Solution: Prepare a solution containing 0.4 mg/mL of USP o-flutamide RS and 0.5 mg/mL of USP flutamide RS in Mobile Phase.

Assay Preparation: Dissolve 50 mg of flutamide, previously dried, in a 100 mL volumetric flask with Mobile Phase and dilute to volume.

Chromatographic System: The chromatography system used is the same as for the Chromatographic Purity procedure. The System Suitability Solution must have a resolution between o-flutamide and flutamide of not less than 2.0. The Standard Preparation must have a relative standard deviation for replicate injections of not more than 1.0%. The Assay Preparation is injected and the peak responses are recorded. The assay value must be between 98.0 to 101.0 percent on a dried basis.

4.2   Elemental Analysis

Flutamide was analyzed for carbon, hydrogen, nitrogen, and fluorine [17]. The carbon, hydrogen, and nitrogen analyses were performed on a Perkin-Elmer Model 240 instrument. Fluorine was determined using an Orion specific ion electrode, and oxygen was calculated by difference. The results obtained during this work are as follows:

Element	Theoretical (%)	Found (%)
C	47.83	47.87
H	4.01	3.75
N	10.14	10.06
F	20.63	20.83
O	17.38	17.49

4.3   Electrochemical Analysis

Differential pulse polarographic assay methods were described [11,13,14] for the quantitative determination of flutamide drug substance and in its formulated products. These methods are based on the reduction of flutamide in aqueous and mixed media to a nitro radical anion, and the corresponding linear relation between the peak current and the flutamide concentration.

4.4   Spectrophotometric Methods of Analysis

When flutamide is dissolved in hydrochloric acid, color formation occurs with an absorption maximum of 380 nm. This reaction was used to develop a simple and sensitive method for the assay determination of flutamide drug substance, and in formulated product [15]. The method was linear over the range 2.5 – 15.0 μg/mL, and the chromophore remained stable for 1 hour in solution.

Flutamide tablets may be assayed using UV spectrophotometry at 304 nm [11]. Tablets are sonicated and diluted in ethanol. After centrifugation, an aliquot is diluted in 20:80 ethanol / 40 mM pH 8.0 Britton Robinson buffer.

4.5   Chromatographic Methods of Analysis

4.5.1   Thin Layer Chromatography

A report [16] summarized the performance characteristics for an over-pressured thin-layer chromatographic analysis (OPTLC) method for flutamide and three related compounds. The report investigated the relationship between various flow rates and the effect on linearity of the calibration graph, the limits of detection and quantitation, and the precision of the data. An additional method for the analysis of flutamide and its related compounds is known [17]. The main characteristics of both methods are as follows:

Stationary Phase	Eluent	Detection Wavelength (nm)	Sample Concentration (mg/mL)	Reference
Silica gel 60F$_{254}$	Chloroform	237	Flutamide (1%) and impurities (10^{-4} to 4×10^{-2} percent) in chloroform	16
Silica gel GF	Chloroform / Methylene chloride (90:15)	254	Flutamide (20 mg/mL) in methanol	17

A TLC method for the detection of the *o*-nitro substituted flutamide isomer, 2-methyl-*N*-(2-nitro-5-trifluoromethylphenyl)propionamide, in impure flutamide was also reported [18].

4.5.2   Gas Chromatography

A gas chromatographic method for the assay of flutamide in tablets has been published [19], and the defining characteristics of this method are as follows:

Instrument	GC with FID and integrator
Column	10% OV-1 on Chromosorb G(AW) treated with DMCS (3 ft x 2 mm id) 80-100 mesh
Carrier Gas	Nitrogen
Flow Rate	40 cm^3/minute
Temperatures:	
Oven	225°C
Injector	275°C
Detector	280°C
Flutamide Standard Stock	5 mg/cm^3 in chloroform (retention time approximately 2 minutes)
Internal Standard	Chlorpheniramine maleate (retention time ~ 3.5 minutes)
Stock solution	10 mg/cm^3 in chloroform
Linearity of Detector Response	1.5-3.0 mg/cm^3

4.5.3   High Performance Liquid Chromatography

In addition to the compendial HPLC methods for purity and assay, an additional method for assay of flutamide drug substance has been developed [17]. The critical test parameters can be summarized as:

Column:	300 x 4 mm µBondapak C18 column
Flow rate:	1 mL/minute
Detection:	UV absorbance at 254 nm
Mobile phase:	7:4 v/v methanol / 50 mM potassium phosphate
Injection volume:	10 µL
Sample solution:	0.16 mg/mL of flutamide containing 0.09 mg/mL testosterone (internal standard) dissolved in methanol
Flutamide retention time:	12 minutes

A recent publication [11] described the following method for assay of flutamide and its impurities in tablets:

Column:	150 x 3.9 mm µBondapak/µPorasil C18 column (operated at 35°C)
Flow rate:	1 mL/minute
Detection:	UV absorbance at 302 nm
Mobile phase:	1:1 v/v methanol / 50 mM phosphate buffer (pH = 3.0)
Injection volume:	20 µL
Sample solution:	1 tablet was sonicated in ethanol and diluted with mobile phase
Retention times:	flutamide (10.0 minutes), 3-trifluoromethyl-4-nitroaniline (4.0 minutes)

4.6   Determination in Body Fluids and Tissues

A number of methods have been reported for the analysis of flutamide in body fluids and tissues. One method for the determination of flutamide and hydroxy-flutamide in dog plasma utilizes mid-bore chromatography [20]. Several advantages were outlined, including reduced mobile phase and sample volumes, no formal extraction process, and adequate accuracy, precision and recovery without the use of an internal standard. This method can also be modified for the analysis of human plasma by simply changing the composition of the mobile phase.

A reverse phase HPLC method was described to measure levels of flutamide and its active metabolite (2-hydroxyflutamide) in rats [21]. The method used a methyltestosterone internal standard, and was described as being sensitive and precise.

Another method reported is an HPLC method with UV detection for the analysis of flutamide, 2-hydroxy-flutamide, and trifluoromethyl-nitroaniline in human plasma [22]. An HPLC method with photodiode array detection and gradient elution was developed for a class-independent drug screen [23], an HPLC method for the plasma analysis of hydroxyflutamide [24], and two GC methods for the analysis of flutamide and its metabolites in human plasma [25, 26].

5.   Stability

5.1   Solid-State Stability

Flutamide shows no loss of potency in the solid state when stored in sealed amber glass bottles at room temperature for up to five years [17].

5.2   Solution-Phase Stability

The stability of flutamide was studied over a wide pH range in 3:2 phosphate buffer / ethanol at refrigerated, room temperature, and 45°C for 1 and 2 weeks [17]. The flutamide concentration was 1.0 mg/mL, and solutions of flutamide in USP ethanol were included as a control. In addition, methanolic solutions of flutamide (0.4 mg/mL) were exposed to 500 foot-candle fluorescent light for up to 8 hours, and then assayed. A summary of the data obtained is presented in Table 8.

Table 8

Solution Stability of Flutamide at Various pH Values

pH	Time (weeks)	Recovery of Flutamide (%)[a]		
		Refrigerated	Room Temp.	45°C
1	1	99 (<0.4)	95 (3.9)	73 (22)
	2	97 (1.0)	89 (8.6)[b]	53 (39)
2	1	101 (<0.4)	100 (<0.4)	99 (2.0)
	2	101 (<0.4)	99 (<0.4)	96 (4.0)
3	1	101 (<0.4)	99 (<0.4)	100 (<0.4)
	2	99 (<0.4)	99 (<0.4)	99 (<0.4)
4	1	100 (<0.4)	100 (<0.4)	100 (<0.4)
	2	101 (<0.4)	101 (<0.4)	101 (<0.4)
5	1	101 (<0.4)	100 (<0.4)	100 (<0.4)
	2	99 (<0.4)	99 (<0.4)	98 (<0.4)
6	1	100 (<0.4)	99 (<0.4)	100 (<0.4)
	2	101 (<0.4)	100 (<0.4)	101 (<0.4)
7	1	100 (<0.4)	100 (<0.4)	100 (<0.4)
	2	101 (<0.4)	101 (<0.4)	101 (<0.4)
8	1	100 (<0.4)	99 (<0.4)	99 (<0.4)
	2	101 (<0.4)	100 (<0.4)	100 (0.7)
10	1	100 (<0.4)	97 (2.1)	82 (15)
	2	100 (0.7)	96 (3.7)	73 (21)
ethanol (control)	1	99 (<0.4)	99 (<0.4)	97 (0.8)
	2	99 (<0.4)	101 (<0.4)	99 (1.1)

a: Estimation of 4-nitro-3-trifluoromethyl-aniline (degradation product) is given in parentheses

b: Another unknown small component was observed

In aqueous alcoholic solutions, flutamide exhibits maximum stability between pH 3 and 8. Under more acidic or alkaline conditions, the major degradation product observed represents hydrolysis of the amide to form 4-nitro-trifluoromethyl-aniline.

Ethanolic solutions (control) showed no degradation for up to 2 weeks when stored at refrigeration and RT, and about 1% degradation at 45°C.

Methanolic solutions (0.4 mg./mL) exposed to 500 foot-candle fluorescent light showed no observable degradation over the 8 hour time period studied.

6.   Drug Metabolism and Pharmacokinetics [27]

6.1   Adsorption

Analysis of plasma, urine, and feces following a single oral 200 mg dose of tritium labeled flutamide to human volunteers showed that the drug is rapidly and completely absorbed. Following a single 250 mg oral dose to normal adult volunteers, the biologically active alpha-hydroxylated metabolite reaches maximum plasma concentrations in about 2 hours, indicating that it is rapidly formed from flutamide.

6.2   Distribution

In male rats, neither flutamide nor any of its metabolites is preferentially accumulated in any tissue (except the prostrate) after an oral 5 mg/kg dose of ^{14}C flutamide. Total drug levels were highest 6 hours after drug administration in all tissues, and levels declined at roughly similar rates to low levels at 18 hours. The major metabolite was present at higher concentrations than flutamide in all tissues studied. Following a single 250 mg oral dose to normal adult volunteers, low plasma levels of flutamide were detected. The plasma half-life for the alpha-hydroxylated metabolite of flutamide is about 6 hours. Flutamide, *in vivo*, at steady-state plasma concentrations of 24 to 78 ng/mL, is 94% to 96% bound to plasma proteins. The active metabolite of flutamide, *in vivo*, at steady-state plasma concentrations of 1556 to 2284 ng/mL, is 92% to 94% bound to plasma proteins.

6.3   Metabolism

The composition of plasma radioactivity, following a single 200 mg oral dose of tritium-labeled flutamide to normal adult volunteers, showed that flutamide is rapidly and extensively metabolized, with flutamide comprising only 2.5% of plasma radioactivity 1 hour after administration. At least 6 metabolites have been identified in plasma. The major plasma metabolite is a biologically active alpha-hydroxylated derivative, which accounts for 23% of the plasma tritium 1 hour after drug administration. The major urinary metabolite is 2-amino-5-nitro-4-(trifluoromethyl)-phenol.

6.4   Elimination

Flutamide and its metabolites are excreted mainly in the urine with only 4.2% of the dose being excreted in the feces over 72 hours.

Acknowledgments

The authors wish to thank the following individuals from the Schering-Plough Research Institute for their assistance: Janet Hatolski for providing the analytical standard used for generation of spectral and other in-house data of flutamide, Rebecca Osterman for obtaining the NMR spectra, Yao Ing for obtaining the mass spectrum, Charles Eckhart for the laser diffraction particle size data, Susan Pasternak for obtaining the DSC/TGA thermograms, Kenneth Cappel for obtaining the x-ray diffraction data, Sudabeh Pakray for literature searches, and Dr. Van Reif for his suggestions and review of this publication.

References

1. *Martindale – The Extra Pharmacopoeia*, 31th edn., J.E.F. Reynolds, ed., The Pharmaceutical Press, London, 1996, p. 575.

2. *Physicians' Desk Reference*, 53rd ed., Medical Economics Company, Montvale, New Jersey, 1999, p. 2840.

3. R. Brogden and S. Clissold, *Drugs*, **38**, 185 (1989).

4. M. Jonler, M. Richmann, and R. Bruskewitz, *Drugs*, **47**, 66 (1994).

5. R. Neri and J. Topliss, Substituted Anilides as Anti-Androgens, **U.S. Patent 4,144,270**, March 13, 1979.

6. L. Peer and J. Mayer, Process for Nitrating Anilides, **U.S. Patent 4,302,599**, November 24, 1981.

7. J. Morris, L. Hughes, A. Glen, and P. Taylor, *J. Med. Chem.*, **34**, 447 (1991).

8. J. Cense, V. Agafonov, R. Ceolin, P. Ladure, and N. Rodier, *Struct. Chem.*, **5**, 79 (1994).

9. R. Ceolin, V. Agafonov, A. Gonthier-Vassal, H. Szwarc, J-M Cense, and P. Ladure, *J. Them. Anal.*, **45**, 1277 (1995).

10. *The Merck Index*, 12th ed., Merck Co., Inc., Whitehouse Station, New Jersey, 1996, p. 713.

11. A. Alvarez-Lueje, C. Pena, L.J. Nunez-Vergara, and J.A. Squella, *Electroanalysis*, **10**, 1043 (1998).

12. Flutamide, *United States Pharmacopoeia 23, Fifth Supplement,* United States Pharmacopoeial Convention, Inc, Rockville, MD, 1996, p. 3416.

13. A. Snycerski, *J. Pharm. Biomed. Anal.*, **7**, 1513 (1989).

14. A. Snycerski and M.K. Kalinowski, *Pol. J. Chem.*, **66**, 49 (1992).

15. S.S. Zarapkar, C.D. Damle, and U.P. Halkar, *Indian Drugs*, **33**, 193 (1996).

16. K. Ferenczi-Fodor and Z. Vegh, *J. Planar Chrom.*, **6**, 256 (1993).

17. Schering-Plough Research Institute, unpublished data.

18. Z. Yang, Y. Xia, Y. Zheng, and P. Xia, *Yaoxue Xuebao*, **29**, 390 (1994).

19. R.T. Sane, M.G. Gangrade, V.V. Bapat, S.R. Surve, and N.L. Chonkar, *Indian Drugs*, **30**, 147 (1993).

20. D. Farthing, D. Sica, I. Fakhry, D. Walters, E. Cefali. and G. Allan, *Biomed. Chrom.*, **8**, 251 (1994).

21. C-J. Xu and D. Li, *Zhongguo Yaoli Xuebao*, **19**, 39 (1998).

22. K. Doser, R. Guserle, R. Kramer, S. Laufer, and K. Lichtenberger, *Arzneim.-Forsch.*, **47**, 213 (1997).

23. O. Drummer, A. Kotsos, and I. McIntyre, *J. Anal. Toxicol.*, **17**, 225 (1993).

24. A. Belanger, M. Giasson, J. Couture, A. Dupont, L. Cusan, and F. Labrie, *Prostate*, **12**, 79 (1988).

25. M. Schulz, A. Schmoldt, F. Donn, and H. Becker, *Eur. J. Clin. Pharmacol.*, **34**, 633 (1988).

26. E. Radwanski, G. Perentesis, S. Symchowicz, and N. Zampaglione, *J. Clin. Pharmacol.*, **29**, 554 (1989).

27. ***Eulexin Capsule Product Information***, copyright 1996, Schering Corporation.

HISTAMINE

Abdullah A. Al-Badr and Hussein I. El-Subbagh

Department of Pharmaceutical Chemistry,
College of Pharmacy
King Saud University,
P.O. Box 2457, Riyadh-11451
Saudi Arabia

Contents

1.   Description [1-3]

1.1   Nomenclature

1.1.1   Chemical Names

2-(Imidazol-4-yl)ethylamine

1*H*-Imidazole-4-ethanamine

2-(4-Imidazolyl)ethylamine

4-Imidazole-ethylamine

5-Imidazole-ethylamine

β-Aminoethylimidazole

β-Aminoethylglyoxaline

4-(2-Aminoethyl)imidazole

1.1.2   Nonproprietary Names

Histamine: Ergotidine

Histamine dihydrochloride: Histamine hydrochloride; Histamini dihydrochloridum

Histamine salicylate: Histamine disalicylate

Histamine diphosphate: Histamine acid phosphate; Histamine phosphate

Histamine picrate: Histamine monopicrate; Histamine dipicrate

1.1.3   Proprietary Names

Histamine base: Histamine, Ergotidine

Histamine dihydrochloride: Amin-glaukosan, Imido, Imadyl, Ergamine, Peremine

Histamine salicylate: Cladene

Histamine diphosphate: Histapon

1.2   Empirical Formulae, Molecular Weights, and CAS Numbers

Histamine	$C_5H_9N_3$	111.5	[51-45-6]
Histamine dihydrochloride	$C_5H_{11}Cl_2N_3$	184.07	[56-72-8]
Histamine salicylate	$C_{12}H_{15}N_3O_3$	249.45	–
Histamine diphosphate	$C_5H_{15}N_3O_8P_2$	307.13	[51-74-1]
Histamine picrate	$C_{11}H_{12}N_6O_7$	340.25	–

1.3 Structural Formulae

1.3.1 Histamine

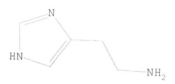

1.3.2 Histamine Diphosphate

$$\left[\begin{array}{c} \underset{3}{\overset{H}{N}} \overset{+}{\underset{1}{N}} \overset{H}{} \\ \end{array} \right] \ 2\ H_2PO_4^{\ominus}$$

1.4 Appearance

Histamine base: Deliquescent needles
Histamine dihydrochloride: Prisms
Histamine salicylate: Crystals
Histamine diphosphate: Prismatic crystals
Histamine picrate: Yellow needles

1.5 Uses and Applications

Since the discovery of histamine in 1910, it has been considered as a local hormone (autocoid), although lacking an endocrine gland in the classical sense. In the past few years, however, the role of histamine as a central neurotransmitter has been recognized. Consequently, a considerable amount of research is now directed toward elucidating its central effects and receptors [4]. The discovery of the duality of the histamine receptor added another dimension to this complex field, leading to new and successful therapeutic, as well as theoretical, investigations [5,6]. Histamine is used as a diagnostic aid (gastric secretion, pheochromocytoma), and in antiallergic (hypersensitization) therapy [1].

2.   Methods of Preparation

Histamine occurs in very small amounts in ergot. It is among the products of the bacterial decomposition of histidine, and this constitutes one of the methods for its production. Histamine is produced from histidine by pneunococci or *E. coli* [10].

Several synthetic procedures have been reported for the preparation of histamine. The drug can be prepared from imidazole propionic acid [7,8], and from α-aminobutyrolactone [9].

3.   Physical Properties

3.1   X-Ray Powder Diffraction Pattern

The X-ray powder diffraction pattern of histamine diphosphate has been obtained using a Philips PW-1710 diffractometer system equipped with a single crystal monochromator and using copper Kα radiation. The powder pattern is shown in Figure 1. The table of scattering angles, d-spacings, and relative intensities were automatically obtained for histamine diphosphate using the Philips digital printer, and are found in Table 1.

3.2   Thermal Methods of analysis

3.2.1   Melting Behavior

The melting point ranges of various histamine salts are as follows:

Histamine base:	83–84°C
Histamine dihydrochloride:	244–246°C
Histamine diphosphate:	140°C
Histamine monopicrate:	160–162°C

3.2.2   Differential Scanning Calorimetry

The differential scanning calorimetry thermogram was obtained for histamine diphosphate using a Dupont TA 9900 thermal analyzer attached to Dupont Data Unit. The thermogram recorded using the purity program is shown in Figure 2. The data were obtained using a heating ramp of 10°C/min, over the range of 40-325°X. It was found that histamine phosphate melted at 134.8°C, and that the associated enthalpy of fusion was 76.0 kJ/mol.

Figure 1.   X-Ray powder diffraction pattern of histamine diphosphate.

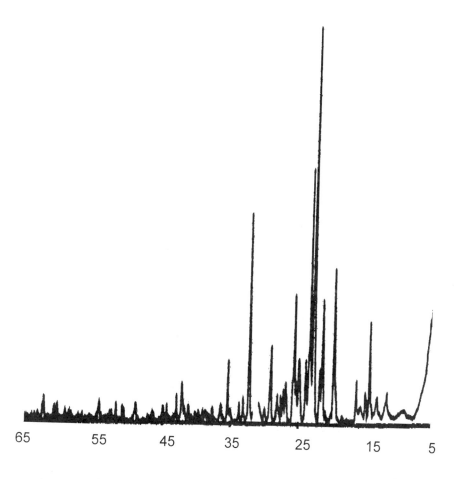

Scattering Angle (degrees 2-θ)

Table 1

Crystallographic Data from the X-Ray Powder Diffraction
Pattern of Histamine Diphosphate

Scattering Angle (deg-2θ)	d-spacing (Å)	Relative Intensity (%)	Scattering Angle (deg-2θ)	d-spacing (Å)	Relative Intensity (%)
5.484	16.1132	27.80	35.335	2.5401	13.77
9.750	9.0716	4.34	36.406	2.4678	4.79
11.821	7.4864	7.02	37.518	2.3972	3.73
13.242	6.6861	7.31	38.550	2.3353	3.22
14.385	6.1572	15.28	39.559	2.2781	2.74
15.034	5.8927	8.98	39.964	2.2559	2.93
15.871	5.5838	16.09	41.052	2.1986	4.66
16.437	5.3930	12.68	41.937	2.1542	10.58
19.678	4.5113	29.61	42.752	2.1150	6.21
21.436	4.1452	30.02	44.165	2.0506	4.05
22.747	3.9091	100.00	44.754	2.0249	4.76
23.261	3.8238	68.49	46.172	1.9600	2.63
23.881	3.7260	14.54	46.861	1.9387	1.41
24.808	3.5889	18.02	48.852	1.8642	4.76
25.631	3.4755	25.78	50.682	1.8011	3.32
26.765	3.3307	9.88	51.713	1.7676	2.92
27.339	3.2620	6.24	52.397	1.7462	2.86
27.946	3.1926	7.08	54.147	1.6938	5.53
29.037	3.0751	20.34	56.952	1.6169	1.99
29.844	2.9937	4.86	58.537	1.5768	2.83
30.639	2.9178	6.18	59.222	1.5602	1.77
31.468	2.8429	4.69	60.322	1.5343	2.63
32.979	2.7 160	5.73	63.052	1.4743	2.48
33.670	2.6618	4.22			

Figure 2.   Differential scanning calorimetry thermogram of histamine
            diphosphate.

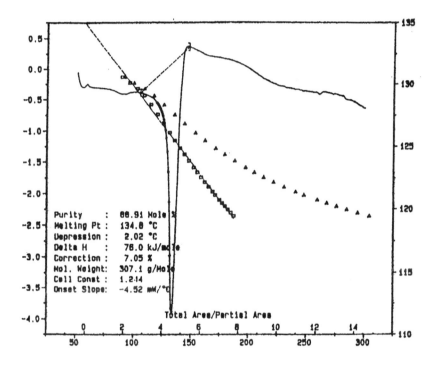

Purity   : 66.91 Mole %
Melting Pt   : 134.8 °C
Depression   : 2.02 °C
Delta H   : 76.0 kJ/mole
Correction   : 7.05 %
Mol. Weight : 307.1 g/Mole
Cell Const   : 1.244
Onset Slope : -4.52 mW/°C

Total Area/Partial Area

Temperature (°C)

3.3   Solubility Characteristics

The following solubility information has been reported for histamine and its salts [1,2]:

> Histamine free base: Freely soluble in water, alcohol, hot chloroform, and sparingly soluble in ether.

> Histamine dihydrochloride: Freely soluble in water, methanol, and ethanol.

> Histamine salicylate: Soluble in water.

> Histamine diphosphate: Readily soluble in water.

3.4   Ionization Constants

Histamine diphosphate was reported to exhibit pKa values of 6.0 and 9.9 [11].

3.5   Spectroscopy

3.5.1   UV/VIS Spectroscopy

The UV-VIS absorption spectrum of histamine diphosphate was recorded on a Shimadzu model 1601 PC Spectrometer. The spectrum of aqueous histamine diphosphate was obtained utilizing matched 1-cm quartz cells, against a corresponding blank. The absorption spectrum is shown in Figure 3, and shows a one band maximum at 211 nm (for which $A_{1\%\text{-}1cm}$ = 153.4, and a molar absorptivity of 4988 L/mole·cm).

Clarke has stated that there is no significant absorption between 230 and 360 nm [2].

3.5.2   Vibrational Spectroscopy

The infrared absorption spectrum of histamine diphosphate was obtained using the KBr pellet method, and was obtained on a Perkin Elmer infrared spectrophotometer. The spectrum obtained is shown in Figure 4, and contained characteristic peaks at energies of 745 and 770 cm^{-1} (CH$_2$), 1460 cm^{-1} (CH), 1490 cm^{-1} (CH$_2$), and 3000-3520 cm^{-1} (NH, NH$_2$).

For histamine hydrochloride in a KBr pellet, Clarke reported principal peaks at energies of 840, 1622, 1522, 1084, 802, and 1577 cm^{-1} [2].

Figure 3. Infrared absorption spectrum of histamine diphosphate.

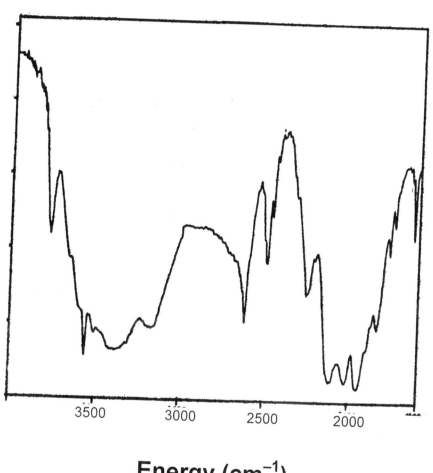

Energy (cm⁻¹)

Figure 4.   Ultraviolet absorption spectrum of histamine diphosphate.

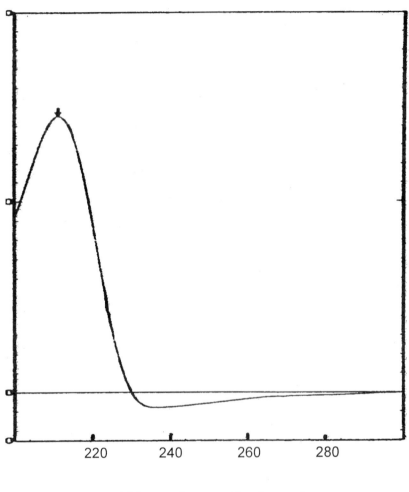

Wavelength (nm)

3.5.3 Nuclear Magnetic Resonance Spectrometry

^{1}H-NMR and ^{13}C-NMR spectra of histamine diphosphate were recorded on a Varian XL 200, 200 MHz spectrometer. The ^{1}H-NMR spectrum was obtained in D_2O solution using tetramethylsilane as the internal standard, and the ^{13}C NMR spectra were obtained in D_2O solution using dioxane as the internal standard. Complete assignments for the observed ^{1}H and ^{13}C NMR resonances are found in Table 2, with explanations following in the next sections.

3.5.3.1 ^{1}H-NMR Spectrum

The ^{1}H-NMR spectrum (Figure 5) shows two heteroaromatic proton singlets at 8.71 and 7.45 ppm that are assigned (on the basis of chemical shift considerations) to the C2 and C5 protons of the imidazole ring, respectively. The two multiplets at $\delta = 3.19$-3.26 and $\delta = 3.37$-3.45 ppm are assigned to the two CH_2 groups of the ethylamine group, at positions 1' and 2', respectively. The assignments were confirmed using the DEPT, COSY, and HETCOR experiments as shown in Figures 6, 7, and 8.

3.5.3.2 ^{13}C-NMR Spectrum

A noise modulated broad band decoupled ^{13}C-NMR spectrum (Figure 9) showed five resonance bands, in accord with 5-carbon structure of histamine. Carbon signals at chemical shifts of 117.93, 129.40, and 134.78 ppm were assigned to the imidazole ring carbons. Signals at $\delta = $ 23.10 and 38.97 ppm were assigned for the ethylamine group carbons atoms. The DEPT experiment (Figure 6) permitted the identification of the methine and methylene carbons at $\delta = 23.10$ and 38.97 ppm as being due to carbons 1' and 2', respectively. The two CH-absorptions at $\delta = $ 117.93 and 134.78 ppm are assigned for carbons 5 and 2, respectively. A two dimensional heteronuclear correlation (HETCOR) experiment (Figure 8) confirmed the afore-mentioned assignments.

3.5.4 Mass Spectrometry

The mass spectrum of histamine diphosphate was obtained utilizing a Shimadzu PQ-5000 mass spectrometer, with the parent ion being collided with helium carrier gas. The mass spectrum is shown in Figure 10 , and Table 3 shows the mass fragmentation pattern.

Clarke listed the principal MS peaks at m/z = 82, 30, 81, 54, 28, 55, 83, and 41 [2].

Figure 5.   ¹H-NMR spectrum of histamine diphosphate.

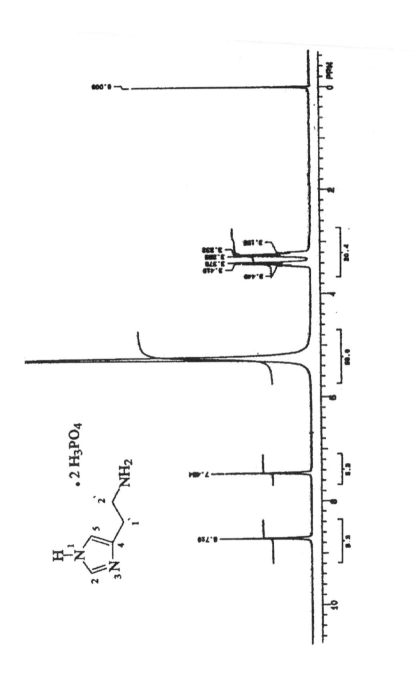

Figure 6.   DEPT-NMR spectrum of histamine diphosphate.

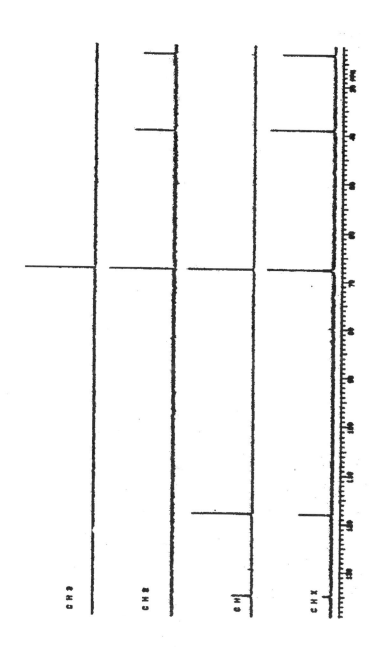

Figure 7.   COSY-NMR spectrum of histamine diphosphate.

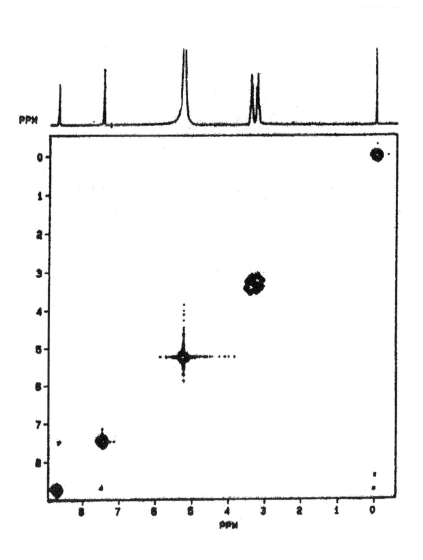

Figure 8.   HETCOR-NMR spectrum of histamine diphosphate.

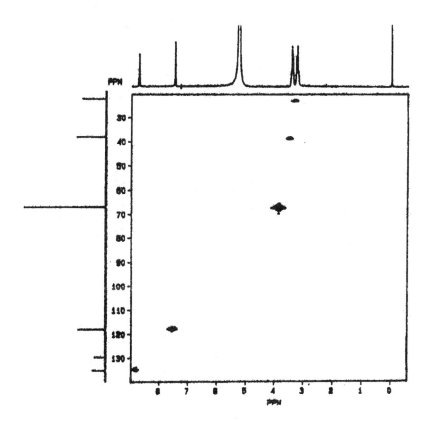

Figure 9. ^{13}C-NMR spectrum of histamine diphosphate.

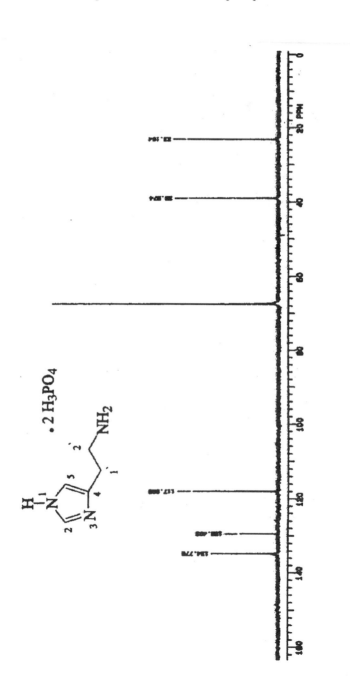

Table 2

^{1}H-NMR and ^{13}C-NMR Assignments for Histamine Diphosphate

^{1}H Assignments

Chemical Shift (ppm, relative to TMS)	Number of protons	Multiplicity (s = singlet, m = multiplet)	Assignment (proton at Carbon #)
3.19-3.26	2	m	1'
3.37-3.45	2	m	2'
7.45	1	s	5
8.71	1	s	2

^{13}C Assignments

Chemical Shift (ppm, relative to TMS)	Assignment (Carbon #)
23.10	1'
38.97	2'
117.93	5
129.40	4
134.78	2

Figure 10.   Mass spectrum of histamine diphosphate.

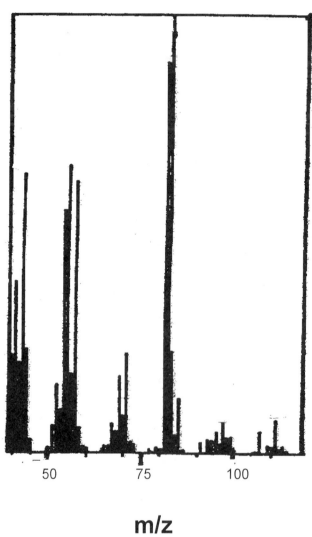

m/z

Table 3

Mass Fragmentation Pattern of Histamine Diphosphate

m/z	Relative intensity	Fragment
111	9%	
97	8%	
85	12%	
82	100%	
81	89%	

Table 3 (continued)

Mass Fragmentation Pattern of Histamine Diphosphate

m/z	Relative intensity	Fragment
57	62%	
56	17%	
55	65%	
54	46%	
43	63%	

4.   Methods of Analysis

4.1   General Reviews

Schwedt presented a review of the methods available for the analysis of histamine and other biogenic amines [19]. Beaven *et al.* have reviewed the past and present assay procedures for the determination of histamine in biological fluids, and considered the clinical application of these methods [20]. Keyzer *et al.* have presented a review on clinical applications of the determination of histamine and histamine metabolite, in urine and blood, and have discussed the normal levels and the clinical applications of monitoring these analytes [21].

Seki has presented a review on separation, detection and determination of histamine, and other amines [22]. A review was presented by Hurst covering extraction, prevention of interference, and determination by high pressure liquid chromatography of histamine and some selected biogenic amines in food [23]. Lorenz and Neugebauer reviewed the current techniques of histamine determination [24].

4.2   Elemental Analysis

The theoretical elemental composition of histamine and its salts are as follows:

	% C	% H	% N	% Cl	% O	% P
Histamine	54.03	8.16	37.81	–	–	–
Histamine dihydrochloride	32.63	6.02	22.83	38.52	–	
Histamine salicylate	57.87	6.06	16.84	–	19.24	–
Histamine diphosphate	19.55	4.92	13.68	–	41.67	20.17
Histamine monopicrate	38.83	3.55	24.70	–	32.91	–

4.3   Identification

The United State Pharmacopoeia describes the following three tests for the identification of histamine phosphate [25]:

Test 1.   Dissolve 0.1 g in a mixture of 7 mL of water and 3 mL of 1 N sodium hydroxide. Add the solution to a mixture of 50 mg of sulfanilic acid, 10 mL of water, 2 drops of hydrochloric acid, and 2 drops of sodium nitrite solution (1 in 10). A deep red color is produced.

Test 2.   Dissolve 50 mg in 5 mL of hot water, add a hot solution of 50 mg of picrolonic acid in 10 mL of alcohol, and allow to crystallize. Filter the crystals with suction, wash with a small amount of ice-cold water, and dry at 105°C for one hour. The crystals obtained melt with decomposition between 250°C and 254°C.

Test 3.   A 1 in 10 solution responds to the tests for phosphates.

The British Pharmacopoeia lists the following four identification tests [26]:

Test 1.   The infrared absorption spectrum is concordant with the spectrum of histamine hydrochloride EP-CRS. Examine as discs prepared using 1 mg of the substances.

Test 2.   The principal spot in the thin layer chromatogram obtained with 1.0% w/v of histamine hydrochloride is similar in position, color, and size to that in the chromatogram obtained with 1.0% w/v of histamine dihydrochloride EP-CRS in water. Silica gel G is the coating substance, and a 75:20:5 v/v mixture of acetonitrile, water, and 13.5M NH_4OH is the mobile phase.

Test 3.   Dissolve 0.1 g in 7 mL of water and add 3 mL of a 20% w/v solution of sodium hydroxide (solution A). Dissolve 50 mg of sulfanilic acid in a mixture of 0.1 mL of hydrochloric acid and 10 mL of water, and add 0.1 mL of sodium nitrite solution (solution B). Add solution B to solution A and mix. A red color is produced.

Test 4.   Dissolve 2 mg of histamine dihydrochloride in 2 mL of water, acidify with 2 M nitric acid, add 0.4 mL of silver nitrate solution, shake, and allow to stand. A curdy, white precipitate is produced.

4.4   Titrimetric Methods of Analysis

4.4.1   Aqueous Titration

The United States Pharmacopoeia 23 [25] recommends use of the following aqueous titration method. Dissolve about 150 mg, accurately weighed, of histamine phosphate in 10 mL of water. Add 5 mL of chloroform, 25 mL of alcohol, and 10 drops of thymolphthalein test solution. The resulting solution is then titrated with 0.2 N sodium hydroxide VS. Each milliliter of 0.2 N NaOH is equivalent to 15.36 mg of $C_5H_9N_3 \cdot 2\ H_3PO_4$.

4.4.2   Non-Aqueous Titration

The British Pharmacopoeia [26] recommends a series of non-aqueous titrimetric methods for the assay of histamine.

Histamine Hydrochloride: Dissolve 80 mg of histamine hydrochloride in 5 mL of anhydrous formic acid, add 20 mL of anhydrous acetic acid, and 6 mL of mercury (II) acetate solution. Titrate with 0.1 M perchloric acid VS and determine the end point potentiometrically. Repeat the operation without histamine hydrochloride, and the difference between the two titrations represents the amount of perchloric acid required for the analyte. Each milliliter of 0.1 M $HClO_4$ VS is equivalent to 9.203 mg of $C_5H_9N_3 \cdot 2\ HCl$.

Histamine Phosphate: Dissolve 0.14 g of histamine phosphate in 5 mL of anhydrous formic acid, and add 20 mL of anhydrous acetic acid. Titrate with 0.1 M-perchloric acid VS and, determine the end point potentiometrically. Repeat the operation without histamine phosphate, and the difference between the two titrations represents the amount of perchloric acid required for the analyte. Each milliliter of 0.1 M $HClO_4$ VS is equivalent to 15.36 mg of $C_5H_9N_3 \cdot 2\ H_3PO_4$.

4.4.3   Potentiometric Methods

Kudoh *et al* reported the contraction of liquid membrane type, histamine ion selective electrode [27]. Karube *et al* determined histamine in freshness testing of meat, with amine oxidase (flavin containing electrode) [28]. A histamine-selective microelectrode, based on a synthetic organic liquid ion exchanger, was described [29]. A liquid ion-exchanger sensitive to histamine was prepared by inserting a 5% (w/w) solution histamine tetrafluorophenyl borate, in 3-nitro-*o*-xylol into one barrel of a two barrel glass electrode, with a tip diameter of 0.1 to 0.5 μm which formed part of the electrochemical cell. The other barrel was filled with aqueous 0.9% sodium chloride. The histamine barrel of the electrode was calibrated at pH 7.4 in 10 μM to 10 mM histamine-free base prepared in various electrolytes and water.

Tatsuma and Watanabe have reported the sensing of histamine and other imidazole derivatives, using haem peptide-modified electrodes [30]. A SnO electrode, covalently modified with haem nonapeptide, was evaluated as an interference-based amperometric sensor for imidazole derivatives in the presence of hydrogen peroxide. Imidazole derivative co-ordinate strongly with Fe(III) in the haem, thus inhibiting catalytic reduction of hydrogen peroxide at the electrode. Electrode response was higher towards imidazole derivatives relative to other compounds, with response times around 10 seconds.

The preparation of a coated-carbon 5 hydroxy tryptamine membrane electrode, based on ion-associating complexes, was described [31]. A coated carbon 5-hydroxy tryptamine membrane electrode was made using the ion associate of sodium tetraphenylborate with 5-hydroxy tryptamine as the active substance, and tri-(2-ethylhexyl) phosphate as a plasticizer. The electrode had a lifespan of 39 days, and selectivity coefficients of histamine and other substances were determined by the mixed solution method.

4.5   Electrochemical Analysis

Jaber used polarography to determine histamine and other imidazole related compounds as their Ni(II) complexes [32]. Electrochemical reduction of the Ni(II) complexes with histamine was studied by differential pulse polarography in 1 M sodium acetate as a supporting electrolyte. Calibration graphs of the peak height of the differential pulse prewave were linear over histamine concentration ranges of 1.5 to 5. μM, and from 10 μM to 0.1 mM.

An electrochemical study of histamine in the presence of metal ions and carbonyl compounds, has been reported [33]. Interaction between Cr(VI) and histamine was quantifiable by impulse differential polarography at - 0.31 and 1.07 V. The interaction between Cr(VI) and histamine was more complex, but also useful for quantitation. Several aldehydes formed electroreducible Schiff bases with histamine, with well defined waves being obtained upon the use of acetaldehyde (-1.366 V), isophthaldehyde (-0.97 V), and acrolein (-1.22 and -1.43 V).

Hidalgo-de Cisneros *et al* described a differential pulse polarography method for the determination of histamine in biological fluids [34]. Samples (20 mL) were added to 5 mL of 0.1 M acetaldehyde, 1 mL of Na_2H_2EDTA, 20 mL of Britton-Robenson buffer, whereupon the pH was adjusted to 8.6 with NaOH. Histamine was determined with the use of a mercury drop electrode, a platinum wire auxiliary electrode, and a Ag-AgCl (KCl) reference electrode. The initial potential was -0.6 V, the drop time was 0.8 seconds, the sweep rate was -6.25 mV/second, and the pulse amplitude was - 50 mV. The calibration graph of peak current (1.248 V) was linear from 1 to 100 µM. Interference from other species was eliminated by the addition of EDTA.

Medyantseva *et al* determined histamine and other biologically active ligands by as a ruthenium complex compound with ruthenium using a voltammetric method [35]. Ruthenium was complexed with histamine, and then the found the ruthenium was determined from the electro-reduction of the complex in a buffer solution (pH 3 – 5).

Adsorptive stripping voltammetric (biological) speciation, of Ni(II) histidine in aqueous ammonia was reported [36]. Voltammograms of mixtures of 0.173 mM $Ni(NO_3)_2$ in ammoniacal buffer solution in the presence of (*L*)-, (*DL*)-histidine, or histamine were obtained at a hanging mercury drop electrode. The potential was scanned from -0.7 to -1.5 V against a Ag-AgCl reference electrode. The effects of buffer concentration and presence of salts (NaCl or KNO_3) were studied.

4.6   Spectrophotometric Methods of Analysis

The interference of scopoletin with the measurement of histamine content in cotton dust has been reported by Fornes *et al* [37]. Xie *et al* described a multivariate spectrophotometry for the qualitative and quantitative analysis of histamine in mixtures [38]. A mathematical model was proposed for the analysis, and using the method, there was no need to acquire a set of sample mixtures with varied concentration. Qualitative and quantitative analysis was accomplished simultaneously with the use of only one sample. This technique had been tested by analyzing a five- and a four-component amino acid mixture using infrared and ultraviolet visible spectrophotometry results. Spectra were comparable with those obtained by computer simulation.

Histamine has been selectively determined in urine after solvent extraction and subsequent Quantitation by atomic absorption spectrophotometry [39]. The authors have also reported a selective determination for histamine in urine by solvent extraction with [3', 3", 5', 5"]-tetrabromophenolphthalein ethyl ester and atomic absorption spectrophotometry [40].

The colorimetric determination of histamine in fish flesh was reported [67]. Yamani and Untermann developed a histidine decarboxylase medium, and reported its application to detect other amino acid decarboxylases [68]. Bacteria were incubated for 48 hours at 30°C in a culture medium containing (L)-histidine, bromcresol green solution, and chlorophenol red solution at pH 5.3. The presence of histidine decarboxylase was indicated by the color change, from light green to violet. The detection limit was 50 mg/L of histamine.

A colorimetric method for the estimation of histamine acid phosphate was reported [69]. The drug was dissolved in 50 mL of 0. 1 N hydrochloric acid, and a 5 mL aliquot was diluted to 50 mL with water. Diazotized 4-hydrazinobenzenesulfonic acid solution (2 mL) was mixed with 5 mL of rectified spirit, the solution was set aside for 2 minutes, and then 5 mL of sample solution was added. After shaking, sitting aside for 5 to 10 minutes, and diluting to 50 mL with phosphate buffer (pH 11.5), the solution was set aside for a further 15 minutes/ The absorbance was then measured at 440 nm.

Yamagami *et al* used colorimetry to assay the enzyme, diamine oxidase (copper containing amine oxidase) [70]. The assay was based on oxidation of histamine, formation of a histamine-Cu(II) cationic complex from the unconsumed substrate, and extraction of the ion by association with tetrabromophenolphthalein ethyl ester into 1,2-dichloroethane. The absorbance was measured at 515 nm.

A colorimetric method for the determination of histamine in mast cells, muscle and urine, by solvent extraction with Cu(II) and tetrabromo-phenolphthalein ester, was reported [71]. Rat mast cells were processed in a medium containing sodium chloride, potassium chloride, disodium hydrogen phosphate, potassium dihydrogen phosphate, calcium chloride, glucose, and heparin, and histamine, was extracted by boiling with water;

muscle was homogenized in $HClO_4$ and centrifuged. Sodium hydroxide was added to the supernatant solution, and histamine was extracted into butanol, flowed by acidification with H_2SO_4. The sample solution containing 5 μg/mL of histamine was mixed with 5 mM $CuCl_2$ and buffer solution (pH 9.5), and allowed to stand for 5 minutes. A 4 mM solution of tetrabromophenolphthalein ethyl ester is ethanol was then added. The cupper histamine-tetrabromophenolphthalein ion associate formed was extracted into 1,2-dichloroethane, with shaking for 10 minutes, and the absorbance of the organic layer was measured at 515 nm.

4.7   Fluorimetric Methods of Analysis

The quantitative fluorimetric determination of histamine in sardines was reported [41]. Wilhelms determined the drug in biological materials by an improved automated fluorimetric method [42]. The drug has also been determined by a bioassay, fluorimetric, and radioenzymic acid methods [43]. Endo reported a simple fluorimetric method for the simultaneous determination of histamine, polyamines, and histone H1, in tissues [44]. Myers *et al* determined histamine in urine by a cation exchange fluorimetric method [45].

Fluorimetric determination, using Bio Rex 70, of peripheral tissue histamine levels in biological tissues has been reported [46]. Luten determined the drug, in canned fish products using an automated fluorimetric method [47]. The detection and quantitation of digestive-tract histamine in rat stomach, using a histochemical phthalaldehyde method and fluorimetric assay, was reported [48].

The fluorimetric determination of histamine in blood, blood plasma, and body tissues, by a simplified method for extraction and isolation of histamine, was described [49]. Simultaneous fluorimetric assay of histamine and histidine in biological fluids using an automated analyzer, was reported [50]. Suhren *et al* have detected and determined histamine and tyramine in milk products by fluorimetry [51]. High-sampling-rate automated continuous flow fluorimetric technique for the analysis of nanogram levels of histamine, in biological samples, was published [52]. Detection and quantitation of histamine in must and wine, was reported [53].

Ramantanis *et al* determined histamine in uncooked sausages by a fluorimetric method, after derivatization with phthalaldehyde and determination by fluorimetry at 445 mm (excitation at 360 nm) [54]. A kinetic fluorimetric determination of histamine and histidine and their mixtures was described [55]. Interference with the fluorimetric histamine assay of biogenic amines and amino acid was also reported [56]. Three mechanisms of interference were detected, identified as mimicking of histamine, suppression of histamine fluorescence, and generation of histamine [56].

Gutierrez *et al* reported a fluorimetric method for the simultaneous determination of histamine and histidine by second derivative synchronous fluorescence spectrometry [57]. The method is based on the formation of fluorescent condensation complexes of histamine and histidine with phthaldehyde in alkaline media and in the presence of 2-mercaptoethanol. The second derivative synchronous fluorescence spectra were recorded by scanning an excitation and an emission monochromator simultaneously from 280 to 400 nm and from 400 to 520 nm, respectively.

The authors have also reported the determination of histamine by derivative synchronous fluorescence spectrometry [58]. Sample solutions containing 20 µg/mL of histamine were mixed with 0.1 sodium hydroxide and ethanolic 0.1% phthalaldehyde (0.3 mL). After 6 minutes, 1M sulfuric acid (0.4 mL) was added, and the mixture was diluted to 10 mL with water. The first and second-derivative synchronous fluorescence spectra were obtained by scanning both monochromators with an 80 nm difference. The excitation monochromator was scanned from 300 to 460 nm, and the emission monochromator from 380 to 450 nm.

The simultaneous determination of histamine and spermidine by second-derivative synchronous fluorimetry was described [59]. Once dissolved, the sample was mixed with 6 mL of 0.1 M NaOH and 0.3 mL of a 0.1% solution of phthalaldehyde in aqueous 20% ethanol. After 6 minutes, 0.4 mL of 1 M H_2SO_4 was added, and the solution was diluted with water to 10 mL. The second derivative fluorescence spectrum of the solution was recorded at 360 to 600 nm (excitation 220 to 460 nm), at a constant wavelength difference of 140 nm. The compounds were determined from the second-derivative peak-to-peak measurements of the maxima and minima fluorescence intensity values (550 and 504 nm, respectively, for histamine, and 386 and 402 nm, respectively, for sperimidine).

Fonberg Broczek *et al* determined histamine in canned fish, using fluorimetric means [60]. Histamine was determined in wines and in other alcoholic beverages by fluorimetry [61]. Siegel *et al* determined histamine in biological materials, and reported their observations on the use of phthalaldehyde condensation for the measurement of histamine [62].

Vidal Carou *et al* determined histamine in fish and meat products by fluorimetry [63]. The selective determination of histamine in fish by flow-injection fluorimetry was reported [64]. Methanolic fish extracts were injected into aqueous hydrochloric acid, and diluted to 6 mM. The solution was mixed at 1 mL/minute with 0.59 mM phthalaldehyde and 0.15 M sodium hydroxide in 1% methanol, and then with 0.37 M H_3PO_4 at 0.9 m/minute at 30°C. The fluorescence was measured at 450 nm (excitation at 365 nm).

A flow-injection, enzyme inhibition, assay method in fish using immobilized diamine oxidase (copper containing amino oxidase) was reported [65]. The sample is mixed with 0.04 M sodium phosphate buffer (pH 7.2), each following at 0.8 mL/minute. Two portions of the mixture are injected, one upstream and one downstream of a reactor containing the enzyme immobilized on CNBr-activated Sepharose. The materials are injected into a stream (0.4 mL/minute) of 0.02 sodium phosphate buffer (pH 7.2). The resulting stream was treated with 0.24 mM phthalaldehyde in 0.3 M NaOH, and, after reaction in a 1 m knitted coil, with 0.37 M H_3PO_4 (each at 0.4 mL/minute). The fluorescence due to histamine in each sample plug was measured at 450 nm (excitation at 365 nm). Microbial and chemical analysis of Chihuahua cheese, and the relationship to histamine and tyramine, was reported [66].

A chemiluminescence determination of histidine and histamine, using catalytically active complexes of Mn(III) with Schiff bases, was reported [74]. The method is based on enhancement of the catalytic activity of Mn(III) in the oxidation of luminol with hydrogen peroxide. To determine 0.1 μM to 1 μM histamine, the test solution was treated successively with 20 mM sulfosalicylaldehyde, 0.2 mM EDTA, 0.1 M borate buffer (pH 9.2), 0.32 mM $MnCl_2$, 80 μM $KMnO_4$, and 0.2 mM luminol (0.25 mL of each). After the mixture was diluted to 4.5 mL with water and treated with 0.5 mL of 20 mM hydrogen peroxide, the luminescence intensity was measured.

Katayama *et al* reported the determination of histamine and other amines by flow-injection analysis, based on aryl oxalate sulforhodamine 101 [75].

A 20 µL portion of histamine was injected into a stream of aqueous 90%
acetonitrile containing 0.02 M hydrogen peroxide. The mixture was
passed at 1.5 mL/minute to a flow cell together with 0.5 mM bis (4-nitro-
2-(3,6,9-trioxadecyloxy-carbonyl)phenyl] oxalate and 0.1 µM
sulforhodamine 101 in acetonitrile. The chemiluminescence was
monitored by a photomultiplier tube, and this method was applied to the
determination of histamine in fish.

4.8   Mass Spectrometric Methods of Analysis

Histamine was determined in urine using chemical ionization mass
spectrometry (CI-MS) [72]. Three ionization methods were compared for
the assay of histamine after formation of N-α-hepta-fluoro-butyryl N-τ
methoxy carbonyl histamine. Of these, chemical ionization at 120 eV
with ammonia as the reagent gas gave the best results, yielding a detection
limit of 4 pg, compared with 13 pg by chemical ionization mass
spectrometry with methane, and 100 pg by electron impact mass
spectrometry (EI-MS) at 70 eV.

Eagles and Edwards reported the determination of histamine and amine
mixtures by fast atom bombardment mass spectrometry (FAB-MS) [73].
Amines were extracted from beef joints or fermented sausages as their
hydrochlorides, and the extracts, together with solutions of individual
amines and their mixtures, were mixed with heptane or hexanediamine
(the internal standard), and dansylated. The dansyl derivatives were
dissolved in acetonitrile, and the solution was concentrated. 2 µL of this
solution was added to 2 µL of glycerol on the fast atom bombardment
copper probe tip. Positive ion spectra were recorded with a Kratos
MS9/50TC mass spectrometer, using fast Xe bombardment (5 to 8 keV).
A linear graph was obtained for histamine.

4.9   Chromatographic Methods of Analysis

4.9.1   Paper Chromatography

Histamine was identified and determined in fish, using a paper
chromatographic method [76].

Clarke [2a] described a system, based on Whatman No. 1 paper, buffered by dipping in a 5% solution of sodium dihydrogen citrate, blotting, and dried at 25°C for 1 hour. The sample consists of 2.5 μL of a 1% solution, dissolved in 2N acetic acid if possible, but otherwise in 2 N HCl, 2 N NaOH, or in ethanol. The developing solvent consists of 4.8 g of citric acid in a mixture of 130 mL of water and 870 mL of *n*-butanol. Visualization is effected using iodoplatinate spray, white, $R_f = 0.00$.

Furuta *et al* separated histamine from other amines in rat kidneys by phosphocellulose chromatography [96]. Kidney was homogenized in $HClO_4$, the homogenate centrifuged, the supernatant solution adjusted to pH 5 to 6 with potassium hydroxide, and the solution centrifuged. The supernatant solution was applied to a phosphocellulose column (3.5 cm x 6 mm) equilibrated with phosphate buffer (pH 6.2). After stepwise washing of the column with phosphate and borate buffer solution, the amines were eluted in a stepwise manner by using a series of borate buffer solutions containing sodium chloride (pH 8). They were derivatized with fluorescamine for the fluorimetric determination.

4.9.2.   Over-Pressured Layer Chromatography

Simon-Sarkadi *et al* reported the determination of histamine and other biogenic amines by personal OPLC [77]. The personal OPLC BS 50 over-pressured layer chromatograph (OPLC-NIT, Budapest, Hungary) was used to determine the biogenic amines in cheese. Amines were extracted with 6% $HClO_4$ and dansylated. Separation was performed on silica gel 60 F_{254} plates (20×20 cm^2) developed with CH_2Cl_2-ethyl acetate (93:7) at 50 bar external pressure, a mobile phase flow rate of 450 μL/min, and a development time of 430 second. Detection was at 313 nm.

Kovacs *et al* separated and quantified by a stepwise gradient of dansylated biogenic amines (including-histamine) in vegetables using personal OPLC [78]. Vegetables were extracted with 6% $HClO_4$ for one hour and derivatized with dansyl chloride. Over-pressured layer chromatography was performed using a BS50 personal OPLC Chromatograph on (20×20 cm^2) 60 F_{254} HPTLC plates sealed on four sides. The dansylated food extracts (5 mL) were applied, and the plates developed with gradient elution (500 mL/min) using hexane-butanol-triethylamine (A, 900:100:91) and hexane-butanol (B, 4:1). The plates were densitometrically scanned at 313 nm.

4.9.3   Thin Layer Chromatography

Several thin layer chromatography (TLC) methods have been used for the determination of histamine in fish products [79], in heparin salts [80], and in feeds and foods [81]. It was also determined as an authentic sample [82], in soy sauce [83], in uncooked sausages [84], in dialyzable materials in allergan extracts [85], in injection of protein hydrolysate [86], as a derivative [87], in the presence of polyamines [88], in wines [89], and the histamine released from rat serosal mast cells [90]. Other reported TLC methods [83-95] are listed in Table 4.

Clarke [2] described a TLC system for the analysis of histamine. The plate is coated with silica gel G, 250 μm thick, dipped in, or sprayed with 0.1 M methanolic KOH and dried. The mobile phase is 100:1.5 methanol: strong ammonia solution. Reference compounds were diazepam (R_f 75), chlorprothixine (R_f 56), codeine (R_f 33) and atropine (R_f 18). The developing agent is acidified iodoplatinate spray, positive.

Clarke [2a] has reported another TLC system: The plates are glass, 20 x 20 cm, coated with a slurry of 30 gm of silica gel G in 60 mL of water to give a layer 0.25 mm thick and dried at 110°C for 1 hour. The sample is 1.0 μL of a 1% solution in 2 N acetic acid. The solvent is 1.5:100 strong ammonia solution / methanol. Prior to the analysis, the solvent is allowed to stand in the tank for 1 hour. Development is effected by ascending in a tank (21 x 21 x 10 cm), the end of the tank being covered with filter paper to assist evaporation (time of run, 30 minutes). The developing agent is acidified iodoplatinate spray, and histamine yields a R_f value of 0.13.

4.9.4   Gas Chromatography

Gas chromatography (GC) was used for the determination of histamine and metabolites [101-105], histamine in food (fish and cheese) [106], in urine using nitrogen-phosphorous detection [107], histamine derivative in urine [108] histamine in meat [109] and urinary N.tau-methylimidazole-4-yl-acetic acid as an index of histamine turnover [110]. Other methods have been reported for histamine and its basic methylated metabolites, in biological materials [111], histamine metabolites in serum, gastric juice or urine [112], and histamine and analogues by ring acetylation [113]. Table 5 summarizes some of these GC methods [108-114].

Table 4

TLC Conditions for the Analysis of Histamine

Support	Solvent system	Visualizing agent and detection	Ref.
Silica gel G	Benzene-triethylamine (5:1)	350 nm irradiation	[83]
Ready plates	Chloroform-methanol-aqueous 25% (ammonia (12:7:1)	Sprayed with methanolic 0.2% 7 chloro-4-nitrobenzo-furan, and evaluated fluorimetrically at 536 nm (excitation at 365 nm)	[84]
Silica gel 60	Chloroform-methanol-25% ammonia-water (29:16:4:1)	Ninhydrin	[85]
Polyamide thin film	Benzene-anhydrous acetic acid (9:1)	Fluorimetric at 450 nm (excitation at 360 nm)	[86]
Alufolien silica gel 60 F_{254}	17 solvent systems	Gaseous HCl	[87]
Alumina 60 F254	Chloroform-triethylamine (4:1)	Densitometric at 400 nm (excitation at 366 nm)	[88]
LHP KDF plates	Acetonitrile-acetone-ammonium hydroxide (3:1:1)	Ninhydrin in acetone-anhydrous acetic acid (1:1, 2 mg/mL). Densitometric at 520 nm.	[89]

Table 4 (continued)

TLC Conditions for the Analysis of Histamine

Support	Solvent system	Visualizing agent and detection	Ref.
Pretreated HP-TLC plates	Benzene-triethylamine (3:1)	Plates were dried and dipped into 20% paraffin in hexane for fluorimetric detection.	[90]
Aluminus sheets coated with 0.2 mm silica gel G-60	Triethylamine-acetone-benzene (1:2:10) and triethylamine-benzene (1:5)	Spectrophotometrically	[91]
Silica .gel G-60	Chloroform-benzene-triethylamine (6:4:1) then benzene-acetone-triethylamine (10:2:1)	At 350 nm.	[92]
Silica gel H	Chloroform-ethanol-methanol-aqueous 17% ammonia (2:2:1:1).	0.5% Ninhydrin and at 520 nm (reference at 695 nm)	[93]
Silica gel 60 F_{254} (20 × 10 cm)	CH_2Cl_2-triethylamine (10:1)	Scanning densitometry at 366 nm.	[94]
Silica gel G 60	Chloroform-benzene-triethylamine (6:4:1)	Ultraviolet at 254 nm.	[95]

Table 5

GC Conditions for the Analysis of Histamine Diphosphate

Column and support	Temperature	Flow Rate carrier gas	Ref.
Fused silica column (25 m × 0.22 mm) with SP 2100 and deactivated with Carbowax 20M	Programmed 160°C to 220°C.	Helium 0.45 mL/min	[108]
WG-11 capillary column (50 m x 0.2 mm) with F.I.D. WG-11 coupled to a mass spectrometer	–	–	[109]
(180 cm × 2 mm) of 3% of SP 2401 or SP 2250 on Supelcoport (100 to 120 mesh) with nitrogen and phosphorus detection).	Programmed from 250°C to 225°C	Helium or Argon-methane (19:1)	[110]
3% of OV 275 with nitrogen-phosphorous detection	170°C	–	[111]
(30 m x 0.32 mm) coated with DB-5, nitrogen and phosphorus detection	Programmed from 90°C to 150°C	Helium	[112]
(2 mL x 2 mm) glass column packed with 3% OV 275 on Chromosorb W HP with nitrogen and phosphorus detection	170°C	Helium 4.0 mL/min	[113]
(180 cm × 2 mm i.d.) packed with 3% OV-225 on Gas Chrom Q (100-120 mesh) at 170°C	170°C	Nitrogen (25 mL/min.) and ^{63}Ni ECD	[114]

Clarke [2] has described a GC system for the analysis of histamine. The column is 2.5 SE-30 on 80-100 mesh Chromosorb G (acid washed and dimethyldichlorsilane-treated), 2 m x 4 mm i.d. glass column (it is essential that the support is fully deactivated). The column temperature is normally between 100°C and 300°C. The carrier gas is nitrogen at 45 mL/min. Suitable reference compounds are *n*-alkanes with an even number of carbon atoms, and the retention index is 1497.

4.9.5   Liquid and High Performance Liquid Chromatography

Liquid chromatography was used for the determination of histamine in the content of non-volatile amines of soybean paste and soy sauce [98]. Histamine was also separated by liquid chromatography, and detected by peroxyoxalate chemiluminescence [99]. Saito *et al* reported a method for the determination of histamine and 1-methyl histamine by liquid chromatography, using on-column derivatization and column-switching techniques [100]. The determination of biogenic amines in beers and their raw materials by ion-pair liquid chromatography with post-column derivatization was described [94].

Several research articles on the analysis of histamine by high performance liquid chromatography (HPLC) have been reported [115-233]. The methods involve the analysis of histamine [115], histamine and its methyl derivative [116,117], histamine in tissues after condensation with phthalaldehyde [118], histamine in fish and fish products [119], elevated histamine concentration in blood plasma [120], detection in fish and fish products [121], histamine in the presence of ethyl ammonium chloride [122] histamine and derivatives [123], histamine in beer [124], histamine in vinegar and fruit juice [125], histamine in wine [126], and fluorescent labeling by derivative [127]. Table 6 lists other reported HPLC methods [128-233].

4.9.6   Affinity Chromatography

The interaction of catechol 2,3-dioxygenase from *Pseudomonas putida* with immobilized histidine and histamine was reported [97]. The peptide was purified from a bacterial crude extract in one step using affinity chromatography on immobilized histamine. Purification was conducted at 20°C, using a jacketed column containing histidine or histamine coupled to Sepharose. 25 mM Tris hydrochloride buffer (pH 7.5) was used as the mobile phase (35 mL/hour), and detection was at 280 nm. The best adsorption and recovery were obtained with histidyl carboxyhexyl Sepharose at pH 6, and histamine carboxyhexyl Sepharose at pH 6-8.

4.9.7 Ion-Exchange Chromatography

Histamine and other amines were determined in cheese and in fish by an automated ion-change chromatographic method [234]. Lewis *et al* have reported a modified method for the isolation and the analysis of brain histamine using Bio-Rex 70 (cation exchange resin) [235].

Andrews and Baldar determined histamine and other polyamines in urine and cell extracts by automated ion-exchange chromatography, using a single mixed Na^+/K^+ buffer system [236]. Histamine was determined in urine, serum, cerebrospinal fluid, and cell extracts with use of an LKB 4400 Amino Acid Analyzer fitted with a column of Ultropac 8 (LKB Biochrom Ltd.). Samples containing large concentrations of ammonia or amino acids were eluted first with 1.2 M sodium citrate buffer (pH 6.45) to remove these compounds, and histamine was then eluted with 0.56 M sodium citrate / 1.6 M potassium citrate buffer (pH 5.60). Detection was by fluorimetry at 425 nm (excitation at 340 nm), or with use of ninhydrin at 440 and 570 nm.

Ion-exchange separation and pulsed amperometric detection was used for the determination of histamine and other biogenic amines in fish products [237]. The homogenized fish was extracted with 18:1:1 aqueous 5% trichloroacetic acid / 6 M HCl / *n*-heptane, and centrifuged at 5000 rpm for 15 minutes. The organic phase was re-extracted with 10 mL of 5% trichloroacetic acid, the aqueous phases combined and diluted to 50 mL with 5% trichloroacetic acid, and cleaned on a LC 18 SPE cartridge. The amine was eluted with 3 M HCl to a volume of 5 mL. A 50 μL portion was analyzed by ion-exchange chromatography on an IonPac CS1O cation-exchange analytical column (with an IonPac CG10 guard column) packed with solvent-compatible polystyrene-divinylbenzene substrate (8.5 μm) agglomerated with 175 nm cation exchange latex for an ion-exchange capacity of 80 μeq per column. Gradient elution with aqueous 90% acetonitrile / 0.5 M perchloric acid / 1 M sodium perchlorate-water, and detection was by integrated pulsed amperometry with an electrochemical detector. The detector consisted of a thin layer electrochemical cell (3.5 μL) equipped with an Au working electrode, stainless steel counter electrode, and combined pH-Ag/AgCl reference electrode.

Table 6

HPLC Conditions for the Analysis of Histamine

Column	Mobile phase and flow rate	Detection	Remarks	Ref.
30 cm x 4 mm Micropak MCH 10	Methanol-phosphate buffer solution (1:1)	Fluorescence	Assay in *Casimiroa edulis* Aqueous or methanolic extract.	[128]
25 cm of RP 18 ultrasphere ODS (5 μM).	Non-linear gradient of methanol in 0.05M sodium acetate-tetrahydrofuran (99:1)	Fluorimetric	Assay in wine	[129]
Stainless steel column (2.5 cm x 4.6 mm) ultrasphere ODS	60 to 80% of methanol-acetonitrile (1:1) in 0.33 mM H_3PO_4 in 16 min. (1.5 to 2.5 mL/min).	At 254 nm	Assay of putrefactive amines in food.	[130]
25 cm × 4.6 mm of Zorbax SCX at 45°C	$0.1M-KH_2PO_4$ buffer (pH 6.1)-methanol (7:3), (1.5 mL/min.).	Fluorimetric at 445 nm (excitation at 359).	Assay in fish by HPLC with post-column reaction.	[131]

Table 6 (continued)

HPLC Conditions for the Analysis of Histamine

Column	Mobile phase and flow rate	Detection	Remarks	Ref.
–	0.1M Phosphate buffer of pH 6.4 containing 0.4% of triethylamine-methanol-acetonitrile (3:2:1)	–	Comparison of a new HPLC method with other methods, in urine.	[132]
15 cm × 4 mm of TSK GEL LS 410 ODS SIL (5 μm)	0.2M Sodium chloride-methanol-0.1M hydrochloride acid (44:48:1) (0.5 mL/min).	Fluorimetric at 450 nm (excitation at 360 nm)	Assay of histamine formed from histidine, in rat stomach and brain.	[133]
25 cm x 4.6 mm of ultrasphere ODS (5 μm)	–	212 nm.	Reversed phase ion pair HPLC assay of the drug and metabolites in rat urine with pentane-1-sulfonic acid as ion pair reagent.	[134]
Nucleosil C8, or Merck RP-8	Acetonitrile-4 mM KH_2PO_4 buffer (pH 3.5) (13:7).	Fluorimetric at 480 nm (excitation at 390 nm).	Assay in tears.	[135]

Table 6 (continued)

HPLC Conditions for the Analysis of Histamine

Column	Mobile phase and flow rate	Detection	Remarks	Ref.
15 cm x 6 mm of TSK IEK 510 SP (5 μm)	0.1M Potassium propionate (pH 4) containing 0.15M potassium chloride	Fluorimetric at 450 nm (excitation at 350 nm)	Assay of histamine methyl transferase	[136]
25 cm x 5 mm of Nucleosil C18 (5 μm)	Sulfuric acid (1 g/L) in aqueous 25% methanol (pH 2.25) (1 mL/min)	Fluorescence at 451 nm (excitation at 353 nm)	Assay of the drug in biological samples after derivatization	[137]
—	—	Fluorescence	Assay in raw fish and fish products	[138]
25 cm x 4 mm of LiChrosorb Si 60 (10 μm).	0.02M Sodium phosphate buffer (pH 2.5) containing 0.02 M sodium pentane sulfonate and 5% of methanol (1 mL/min)	Fluorescence at 450 nm (excitation at 365 nm)	Assay in blood by HPLC with reaction of the effluent with phthaldehyde.	[139]

Table 6 (continued)

HPLC Conditions for the Analysis of Histamine

Column	Mobile phase and flow rate	Detection	Remarks	Ref.
30 cm x 3.9 mm of μBondapak C18 (10 μm)	Methanol-0.8 acetic acid-acetonitrile (52:45:3) and 100% methanol.	Fluorimetric at 440 nm (excitation at 340 nm)	Assay in must and wine.	[140]
Stainless steel 30 cm x 4 mm μBondapak C18	0.2M of butane-pentane-hexane-octane or (+)-camphor sulfonate (as sodium salt) in methanol-water-acetonitrile (3:17:2) or acetonitrile-water (1:4), at pH 3.	220 nm.	Assay of histamine in cheese.	[141]
10 cm Axxion C18 (3 μm)	6.3% Acetonitrile and 2.3% methanol in sodium acetate solution of pH 4.8.	Electro-chemical (ESA coulochem)	Drug in cerebrospinal fluid was derivatized with 60 mM succinimidyl propionate	[142]

Table 6 (continued)

HPLC Conditions for the Analysis of Histamine

Column	Mobile phase and flow rate	Detection	Remarks	Ref.
10 cm x 6 mm TSK gel SP-2SW (5 μm)	0.25M KH$_2$PO$_4$ (0.6 mL/min.).	Fluorimetry at 450 nm (excitation at 360 nm)	Analysis of plasma in brain. Cation-exchange HPLC coupled with post column derivatization fluorimetry.	[143]
10 cm × 4.6 mm Spherisorb ODS 3 (3 μm)	0.2 mM Sodium acetate-tetrahydrofuran (3:1) adjusted to pH 5.1.	Fluorimetry at 430 nm (excitation at 350 nm)	Highly sensitive HPLC technique for simultaneous measurement of the histamine, methyl-histamine and others.	[144]
RP 8 (10 μm) equipped with precolumn of RP 8 (20 to 40 μm)	Acetonitrile-4 mM phosphate buffer of pH 3.5 (13:7), (1 mL/min.).	Fluorimetric at 480 nm (excitation at 390 nm)	Analysis in tears	[145]
10 μm RP 8 column connected to a precolumn of RP 8 (20 to 40 μm)	4 mM Phosphate buffer (pH 3.5)-acetonitrile (7:13) (0.5 mL/min).	Fluorimetric at 480 (excitation at 390 nm)	Rapid HPLC for determining level of histamine in tears from normal and inflamed human eyes.	[146]

Table 6 (continued)

HPLC Conditions for the Analysis of Histamine

Spherisorb ODS (5 μm)	1M Hydrochloric acid	Fluorimetric	Analysis in food (cheese, fish, yogurt, chocolate or beverages)	[147]
15 cm x 4.6 mm of Nucleosil C18 (5 μm)	5 mM Pentane sulfonic acid in 15 mM citrate buffer containing 5% of methanol, (0.4 mL/min.).	Electro-chemical and fluorimetric at 418 nm (excitation at 350 nm)	Analysis in rat tissues and blood samples	[148]
15 cm x 4.6 mm of Nucleosil C18 (5 μm)	0.07M Na_2HPO_4-citric acid buffer of pH 4.5 containing 23% of acetonitrile and 0.26 mM-Na_2EDTA (1 mL/min.).	amperometry	Analysis of histamine, in brain, as the iso-indole derivative	[149]

Table 6 (continued)

HPLC Conditions for the Analysis of Histamine

Column	Mobile phase and flow rate	Detection	Remarks	Ref.
5 cm x 4 mm of Hitachi gel 2619 F at 70°C with precolumn of μBondapak C18	Trisodium citrate dihydrate (27.5 g)-sodium chloride (117 g), 35% hydrochloric acid (5.26 mL), Brij 35 (0.2 g), aqueous. 10% Na_2EDTA dihydrate (3.7 mL), water (800 mL) and methanol (250 mL).	Fluorimetric at 418 nm (excitation at 370 nm).	Direct and sensitive method. Analysis in acid-deproteinized biological samples (brain and stomach tissues).	[150]
15 cm x 4.6 mm of Nucleosil C18 (5 μm)	0.07M Na_2HPO_4-citric acid buffer solution (pH 4.5) containing 26% of acetonitrile and 0.26 mM-Na_2EDTA (0.8 mL/min.).	amperometry	Analysis of histamine in plasma	[151]

Table 6 (continued)

HPLC Conditions for the Analysis of Histamine

Column	Mobile phase and flow rate	Detection	Remarks	Ref.
9 cm x 4.5 mm of BC X-12 cation exchange resin in an analyzer	0.4 Lithium citrate-0.12M KCl-0.3M H_3BO_3 containing 2.5% of ethanol (pH 8.5.).	Fluorimetric	Polyamine analysis in Picea needles, without previous extract purification. Liquimat Labotron amino acid analyzer	[152]
25 cm x 4.6 mm of Ultrasphere ODS and Whatman CSKI guard column of μBondapak C18 Corasil	0.14M sodium acetate-methanol (17:73), 3.89 mM in octane-1-sulfuric acid and containing EDTA at pH 3.48 (1 mL/min.).	Electro-chemical	Analysis of rat brain simultaneous analysis of the drug and derivative	[153]
Column of Bio-Rad HPX-72-0	3 mM Sodium hydroxide	At 208 and 214 nm.	Analysis of the drug and other food-related amines.	[154]
22 cm × 4.6 mm of Kontron RP 18 (5 μm)	Aqueous 40% acetonitrile containing 50 mM NaH_2PO_4 (0.7 mL/min.).	Fluorimetric at 450 nm (excitation at 350 nm).	Analysis in fish by reversed-phase HPLC	[155]

Table 6 (continued)

HPLC Conditions for the Analysis of Histamine

Column	Mobile phase and flow rate	Detection	Remarks	Ref.
30 cm x 3.9 mm of µBondapak C18	0.M Phosphate buffer-methanol (3:1) containing 3.2 mL/l of 10 M sodium hydroxide and 56 mg/l of sodium dodecyl sulfate; (1.2 mL/min.).	At 208 nm and electro-chemically.	Analysis in biological fluid of the drug and derivative, the method is simple and sensitive.	[156]
25 cm x 4.6 mm of CP sphere C18 with guard column (7.5 cm × 2.1 mm.)	10 mM Sodium heptane-sulfonate/10 mM KH_2PO_4 (pH 3)-methanol (17:3) (1.5 mL/min.).	At 215 and 260 nm.	Analysis of aromatic biogenic amines and their precursors in cheese, guard column of pellicular reversed phase material.	[157]
20 cm × 2.1 mm of Hypersil ODS (5 µm)	25 mM KH_2PO_4-acetonitrile (1:1) (0.5 mL/min.).	Fluorimetric at 447 nm (excitation at 358 nm)	Analysis in fish for the drug and its derivatives.	[158]

Table 6 (continued)

HPLC Conditions for the Analysis of Histamine

Column	Mobile phase and flow rate	Detection	Remarks	Ref.
25 cm x 4.6 mm of Ultrasphere ODS (5 μm)	Methanol-acetonitrile-60 mM KH_2PO_4 of pH 6.4 (3:2:5), (0.7 mL/min.).	Fluorimetric at 375 nm (excitation at 310 nm).	Rat muscles, brain, stomach and heart.	[159]
25 cm x 4.6 mm of Nucleosil C18 (5 μm)	A gradient between methanol-80 mM-acetic acid-acetonitrile (9:8:3) and methanol	Fluorimetric	Detection of the drug and polyamines formed by clostridial putrefaction of caseins	[160]
10 cm x 2.1 mm of Hypersil ODS (5 μm)	0.02M-acetic acid-acetonitrile - methanol (2:9:9) in acetonitrile-0.02M acetic acid (1:9) (0.7 mL/min.).	At 254 nm	Analysis of biogenic amines by RP-HPLC for monitoring microbial spoilage of poultry	[161]

Table 6 (continued)

HPLC Conditions for the Analysis of Histamine

Column	Mobile phase and flow rate	Detection	Remarks	Ref.
15 cm x 4.6 mm of Octylsilane (5 μm)	40 to 10% of aqueous 5 mM-tetrabutyl-ammonium phosphate in methanol-acetonitrile (1:1) (1.5 mL/min.).	At 254 nm	Analysis of amines and their amino acid precursors in protein foods	[162]
25 cm x 4 mm of Nucleosil C18 (5 μm)	5% Methanol in 0.2M sodium chloride-methanol (9:11), adjusted to pH 3 (0.5 mL/min.).	At 450 nm (excitation at 350 nm)	Analysis of plasma histamine after precolumn derivatization.	[163]
25 cm x 3 mm of LiChrosorb RP 8 (10 μm)	Gradient elution from aqueous 70 to 80% methanol, over 7 min., then to 100% methanol over 4.5 min. and pack to aqueous 70% methanol in 1 min.	At 254 nm	Simultaneous method for assay of histamine and other amines, in fish, after derivatization.	[164]

Table 6 (continued)

HPLC Conditions for the Analysis of Histamine

Column	Mobile phase and flow rate	Detection	Remarks	Ref.
15 cm x 4.6 mm of Spherisorb 5C6 (5 μm)	Isocratic elution with methanol-0.08 LiClO$_4$ [14:11 (A) or 7:3] or gradient elution with A (held for 10 min.) containing 0 to 30% of methanol over 40 min and to 40% of methanol over 10 min.	By a 5100 A ESA coulochem detector	Analysis of catabolic amines in silage	[165]
25 cm x 4.5 mm of Ultrasphere Si	Ethyl acetate-hexane-ethanolamine (60:40:0.01).	Fluorimetric at 470 nm (excitation at 335 nm)	Analysis of biogenic amines in fish and fish products	[166]
Brownlee RP 18	Aqueous 90 or 60% acetonitrile containing 75 mM NaH$_2$PO$_4$	–	Analysis in fish by RP-HPLC	[167]

Table 6 (continued)

HPLC Conditions for the Analysis of Histamine

Column	Mobile phase and flow rate	Detection	Remarks	Ref.
25 cm × 4.6 mm of Altex C8 (5 μm)	Acetate buffer (pH 4.6) consisting of anhydrous 0.6% acetic acid, 20% of acetonitrile and 0.1% of pentane-sulfonic acid (1.2 mL/min.).	Fluorimetric at 430 nm (excitation at 365 nm).	Rapid analysis of the drug in plasma and cellular components of blood.	[168]
Shim pack CLC ODS at 50°C	A gradient elution with 0.1M NaH_2PO_4-10 mM sodium octane sulfonate at pH 4.28 (A) in A-methanol (1:1) at pH 4.2.	Fluorimetric at 450 nm (excitation at 348 nm).	Analysis of polyamines in red meat fish and evaluation of freshness.	[169]
15 cm x 3.9 mm of μBondapak CN	Methanol-water-tetrahydrofuran (9:9:2) containing 5 mM octane sulfonic acid pH 3.5 (1 mL/min.).	Fluorimetric at 450 nm (excitation at 360 nm)	Evaluation of interference in basophil histamine release measurement.	[170]

Table 6 (continued)

HPLC Conditions for the Analysis of Histamine

Column	Mobile phase and flow rate	Detection	Remarks	Ref.
20 cm x 4.6 mm of Hypersil ODS (5 μm) at 35°C.	Gradient elution with ethanol-acetonitrile-buffer (pH 8) (1.4 mL/min.).	At 254 nm and at 485 nm (excitation at 245 nm).	Analysis of biogenic amines in cheese.	[171]
15 cm x 3 mm or 12 cm x 4 mm of LiChrospher RP 18	Ethanol-acetonitrile-tris-acetic acid buffer (pH 8) (0.7 mL/min.).	At 254 nm and at 490 nm (excitation at 360 nm).	Analysis of biogenic amines in foods.	[172]
15 cm x 6 mm of Shim Pak CLC-ODS, at 50°C.	Gradient with (i) 0.01M sodium hexanesulfonate in 0.1M $NaClO_4$ (pH 4) and (ii) a mixture of (i) and methanol (1:3) adjusted to pH 3.0, (1.1 mL/min.).	Fluorimetric at 455 nm (excitation at 345 nm)	Simultaneous analysis of amines by RP-HPLC.	[173]
5 cm x 4 mm of Shim Pak ISC 05/SO504P, at 70°C.	Linear gradient from 0.7 to 2.5 M sodium chloride in 0.2N sodium citrate, (0.6 mL/min.).	Fluorimetric at 430 nm (excitation at 345 nm)	Study of vital reaction in wounded skin.	[174]

Table 6 (continued)

HPLC Conditions for the Analysis of Histamine

Column	Mobile phase and flow rate	Detection	Remarks	Ref.
10 cm x 4.6 mm of Phenyl Spheri-5 (5 μm Brownlee) with a precolumn (1.5 cm × 3.2 mm) of New Guard Aquaphore silica (5 μm Brownlee).	Methanol-20 mM sodium acetate buffer (7:3), 0.33 mM sodium octane-1-sulfonate was the ion-pairing agent.	Fluorimetry at 450 nm (excitation at 350 nm)	Analysis in challenged human leukocyte preparations.	[175]
20 cm x 4.6 mm of Spherisorb ODS 2 (5 μm)	0.1 M sodium phosphate buffer of pH 6.5-0.19 mM sodium dodecyl sulfate containing 25% of methanol (1.2 mL/min.)	Electro-chemical	Separation and analysis of the drug and derivatives in brain extracts.	[176]
25 cm x 4.6 mm of Supelcosil LC (18.5 μm) or:	Phosphate buffer solution of pH 3-acetonitrile (7:3) (0.7 to 0.8 mL/min).	Fluorimetric at 470 nm (excitation at 375 nm).	Analysis in fish products	[177]

Table 6 (continued)

HPLC Conditions for the Analysis of Histamine

Column	Mobile phase and flow rate	Detection	Remarks	Ref.
15 cm x 4.6 mm of Hypersil 5 MOS	As above (0.6 to 0.7 mL/min.).	At 285 nm.	–	
25 cm × 4.6 mm of Spherisorb ODS 2 (5 μm) and precolumn (3 cm × 3.9 mm of Bondapak-C_{18} / Corasil (37 to 50 m particle size).	Gradient in 0.05M sodium acetate-0.25 acetonitrile-0.05 triethyl-amine - 1 ppm EDTA (pH 6.8) and acetonitrile, 1 mL/min., at 52°C. Sodium acetate pH (6.8)-acetonitrile	Diode-array detection at 254 nm	Analysis in wine precolumn derivatization. Analysis of histamine and amino acids.	[178]
25 cm x 4.6 mm of Spherisorb ODS 2	5 mM octyl ammonium salicylate or 5 mM octyl ammonium dihydrogen phosphate as interaction reagent.	Ultraviolet	Ion interaction reversed-phase HPLC. Analysis of food related amines.	[179]

Table 6 (continued)

HPLC Conditions for the Analysis of Histamine

Column	Mobile phase and flow rate	Detection	Remarks	Ref.
2 cm x 4 mm of Amberlite CG-50 (Na^+, 200 to 400 mesh)	0.5M hydrochloric acid, the eluent was analyzed by the following column: [(15 cm x 4 mm) of TSK gel CM_2SW (5 μm)] with citric acid-imidazole in H_2O-acetonitrile (4:1) (1 mL/min)	Fluorimetric at 410 to 800 nm (excitation at 230 nm to 400 nm).	Assay of histamine N-methyl-transferase in rat kidney extract.	[180]
25 cm x 4 mm LiChrospher ODS (5 μm)	0.5% Ammonia in aqueous 50% acetonitrile (1.5 mL/min.).	254 nm.	Analysis in fish with offline derivatization.	[181]
25 cm x 4.5 mm of Zorbax ODS C18 (5 μm)	Phosphate buffer (pH 5.6)-aqueous 50% methanol (79:21).	amperometry	Selection of solvent system for the separation and quantitation in plasma.	[182]

Table 6 (continued)

HPLC Conditions for the Analysis of Histamine

Column	Mobile phase and flow rate	Detection	Remarks	Ref.
25 cm x 4.5 mm of Zorbax ODS C18 (5 μm)	Phosphate buffer solution (0.12 M NaH$_2$PO$_4$, 0.1M sodium hydroxide and 19 μM Lauryl sulfate)-methanol (79:21; pH 5.6).	amperometry	Rapid assay method for histamine in small plasma volumes.	[183]
15 cm x 3 mm of SGX C18 (5 μm)	Aqueous 55% methanol (0.5 mL/min) and 0.5 mL of 11.5 ppm of heptane-1,7-diamine (internal standard) and methanol-1,6-dioxane-water (45:12:43).	At 254 nm	Assay of amines in silage	[184]
20 cm x 4.5 mm of RP-18 (5 μm)	Acetonitrile-octan-2-ol in 0.5% Na$_2$HPO$_4$ (1 mL/min.).	Fluorimetric at 445 nm (excitation at 356 nm).	Analysis of amines in wine.	[185]

Table 6 (continued)

HPLC Conditions for the Analysis of Histamine

Column	Mobile phase and flow rate	Detection	Remarks	Ref.
12.5 cm × 4 mm of LiChrospher 100 RP-18	Isocratic elution with 55% methanol or gradient elution with methanol-water (1.5 mL/min.).	At 254 nm	Simultaneous analysis of amines in canned fish.	[186]
10 cm x 4.6 mm of Partisil 5 SCX RAC-II (5 μm)	Si sat'd. 0.17M KH_2PO_4 (1 mL/min.).	Fluorimetric at 418 nm (extinction at 230 nm).	Analysis in rat bronchoalveolar lavage fluid by HP-CEC coupled with post column derivatization.	[187]
15 cm x 4.6 mm Spherisorb 3S TG (3 μm) and a Spherisorb 5 ODS 2 guard column	Acetonitrile, 0.01 M K_2HPO_4, buffer solution (pH 7) and water (0.8 mL/min.).	At 254 nm.	Analysis of amine in food. Improvement of extraction procedures.	[188]

Table 6 (continued)

HPLC Conditions for the Analysis of Histamine

Column	Mobile phase and flow rate	Detection	Remarks	Ref.
15 cm x 4.6 mm of Asahipak ODP 50	50 mM 41:9 $Na_2B_4O_7$-acetonitrile, with 1 mM o-phthaldehyde and 1 mM N-acetyl-L-cysteine (0.5 mL/min).	Fluorimetric at 450 nm (excitation at 340 nm)	Analysis of the drug and derivative, in food, with on column fluorescence derivatization.	[189]
25 cm x 4.6 mm of Nucleosil 7C6H5	25 mM NaH_2PO_4-15% acetonitrile and 25 mM NaH_2PO_4-25% acetonitrile.	Fluorimetric at 450 nm (excitation at 350 nm).	Analysis of the drug in food, combined ion-pair extraction and RP-HPLC	[190]
20 cm x 4.6 mm of Nucleosil 100-5C18	30 to 80% acetonitrile in 0.08 acetic acid (1 mL/min.).	Fluorimetric at 440 nm (excitation at 230 nm).	Analysis of amines in wine using automated precolumn derivatization.	[191]
15 cm x 4.6 mm of Supelco LC ABZ	Phosphate buffer solution (pH 7.3)/ acetonitrile (9:1) (1.2 mL/min.).	At 210 nm (acquisition from 195 to 230 nm).	Analysis of preserved fish product directly.	[192]

Table 6 (continued)

HPLC Conditions for the Analysis of Histamine

Column	Mobile phase and flow rate	Detection	Remarks	Ref.
30 cm x 3.9 mm of µBondapak C18	0.2M potassium hydrogen phthalate pH 3-methanol (9:11) (0.8 mL/min.).	Fluorimetric at 420 nm (excitation at 244 nm).	Analysis of drug in human brain by RP-HPLC.	[193]
25 cm x 4.5 mm of Zobrax ODS C18 stainless steel.	0.12M Phosphate buffer (pH 5.6)-50% methanol (79:21)	amperometry	Application of a novel technique for analysis of the drug content of urine, brain and cerebrospinal fluid.	[194]
–	–	Fluorescence	Ion-pair extraction of the drug from biological fluids and tissues for analysis by RP-HPLC.	[195]

Table 6 (continued)

HPLC Conditions for the Analysis of Histamine

Column	Mobile phase and flow rate	Detection	Remarks	Ref.
25 cm x 4 mm of Chromopak C18 (5 μm) at 20°C.	Water-acetonitrile-triethylamine-H_3PO_4 (425:75:2:2) adjusted to pH 6.4 with ammonium hydroxide (0.7 mL/min.)	Fluorimetric at 440 nm (excitation at 360 nm).	Analysis of the drug in biological samples, rapid and highly sensitive.	[196]
–	–	Fluorimetric	Simultaneous analysis of histamine secretion and calcium in rat basophilic leukemia cells.	[197]
25 cm x 4.6 mm on Spherisorb C_{18} at 25°C	Methanol-acetate buffer solution pH 4.5 (9:11) (0.7 mL/min).	Fluorescence at 445.5 nm (excitation at 350 nm)	Analysis of histamine in platelets with precolumn derivatization with o-phthaldehyde.	[198]

Table 6 (continued)

HPLC Conditions for the Analysis of Histamine

Column	Mobile phase and flow rate	Detection	Remarks	Ref.
25 cm x 4 mm 7 μm Zorbax ODS	Methanol-acetonitrile-water (25:1:14) containing 10 mM-hydrogen peroxide (1 mL/min)	Chemi-luminescence	Analysis of polyamines in plants	[199]
8 cm x 4.6 mm at 37°C on a PTFE-lined ODS (3 μm)	0.1 M sodium acetate containing 25% of acetonitrile and 30% of THF, pH (6.5) (0.8 mL/min).	Electrochem. coulometric	Analysis of amines in wine with pre-column derivatization with phthaldehyde.	[200]
15 cm x 6 mm of TSK gel Catecholpak at 40°C	0.2-Propionate buffer (pH 4.5) containing 86 mM-sodium chloride (1 mL/min)	Fluorescence at 430 nm (excitation at 360 nm)	Analysis of histamine in serum and urine	[201]
25 cm x 4.6 mm of Spherisorb ODS-2 (5 μm) at 60°C with a pellicular C_{18} guard column at 60°C	Aqueous 25-80% acetonitrile over 55 minutes (1 mL/min)	UV at 250 nm	Analysis of biogenic amines in wines	[202]

Table 6 (continued)

HPLC Conditions for the Analysis of Histamine

Column	Mobile phase and flow rate	Detection	Remarks	Ref.
25 cm x 4.6 mm of Cosmosil 5 C_{18}-AR (5 μm)	Aqueous 65% acetonitrile (1 mL/min)	Fluorescence at 525 nm (excitation at 325 nm)	Analysis of amines in food following sample clean up by solid-phase extraction	[203]
Hamilton PRP-X200 No. 79441 Cataionic or: LiChrospher RP-18 (5 μm)	0.4 M-Potassium dihydrogen phosphate of pH 4.5 (0.5 mL/min) or 0.05-sodium dihydrogen phosphate (0.5 mL/min) acetonitrile (7:3) of pH 4.5	UV at 210 nm fluorimetric at 430-470 nm (excitation at 305-395 nm).	Analysis of histamine in fish tissues.	[204]
25 cm x 4.6 m Phenomenex Ultramex C-8 (5 μm) equipped with a Pelligund LC-8 guard column	Acetonitrile – 0.05 M imidazole nitrate buffer of pH 7 (1:4) (1 mL/min).	Chemi-luminescence or fluorescence at 440 nm (excitation at 366 nm)	Analysis of the fluorescence derivative of histamine. Comparison of fluorescence versus chemi-luminescence detection	[205]

Table 6 (continued)

HPLC Conditions for the Analysis of Histamine

Column	Mobile phase and flow rate	Detection	Remarks	Ref.
25 cm x 4 mm i.d. of Spherisorb ODS-2 (5 μm) with a Spherisorb ODS-2 guard column.	Methanolic 50% acetonitrile-0.5 mM H_3PO_4 [from 60:40 to 80:20 (held for 5 min) in 15 min then to 95:5 (held for 5 min) in 5 min]	UV at 254 nm	Analysis of biogenic amines in table olives	[208]
(5 μm) 25 cm x 4.6 mm i.d. Spherisorb ODS column protected by a precolumn (20 cm × 4 mm i.d.) of the same material.	Methanol-0.02 sodium acetate (1:1) supplemented with 5 mM-1-octane sulfonic acid (1 mL/min)	Fluorimetry at 455 nm (excitation at 360 nm).	Analysis of histamines in mast cells	[209]
7 μm reversed phase Nucleosil C8 (25 cm x 4 mm i.d.)	Methanol-aqueous 50 mM NaH_2PO_4 buffer (pH 3.1) containing 0.5 mM Na_4EDTA and 5 mM-pentane-1-sulfonic acid (3:17) (1 mL/min).	UV at 215 nm	Separation and analysis of histamine and its metabolites applied to biological samples.	[210]

Table 6 (continued)

HPLC Conditions for the Analysis of Histamine

Column	Mobile phase and flow rate	Detection	Remarks	Ref.
20 cm x 3 mm Inertsil ODS-2 (5 μm)	H_2O-acetonitrile-methanol-H_3PO_4 (375:75:50:1) of pH 6.87 (0.4 mL/min).	–	Analysis of histamine in biological fluids using fluorescamine as derivitizing agent	[206]
25 cm x 4.6 mm i.d.) of Phenomenex Ultramex C8 (5 μm) with a guard column packed with Pelliguard LC-8	Acetonitrile-0.5 M imidazole buffer of pH 7 (1:4) (1 mL/min)	Fluorimetry at 440 nm (excitation at 366 nm)	Analysis of histamines in plasma using solid-phase extraction and fluorescamine derivatization	[207]
25 cm × 4 mm i.d. 4 μm of Superspher 100RP-18.	Acetonitrile-sodium acetate buffer of pH 6.9 containing acetonitrile and THF (100:10:1) [3:7]	Fluorimetry at 450 nm (excitation at 350 nm)	Analysis of small amounts of histamine in wine and sparkling wine	[211]
15 cm × 4.6 mm i.d. of TSK-gel ODS 80Ts	0.1 M phosphate buffer (pH 6.4) (0.4 or 0.7 mL/min)	Chemi-luminescence	Analysis of histamine in rat tissues. HPLC coupled with immobilized diamine oxidase	[212]

Table 6 (continued)

HPLC Conditions for the Analysis of Histamine

Column	Mobile phase and flow rate	Detection	Remarks	Ref.
10 cm × 3.2 mm i.d., (3 μm) phase II ODS	0.1 M sodium phosphate buffer containing 0.4% triethylamine (pH 6.4), 16% methanol, 14% acetonitrile and 1 mM Na$_2$EDTA.	Electro-chemical	Assay of histamine in mast cells	[213]
JASCO 860-CO at 40°C	5% ethanol in 150 mM sodium acetate buffer of pH 6 and 60% of acetonitrile in water (1 mL/min)	Colorimetric at 420 nm	Analysis of histamine in fish, method is suitable for other imidazole compounds	[214]
40 cm × 27 or 40 mm i.d., (5 μm) YMC-ODS-AQ AQ column	75 mM-H$_3$PO$_4$-0.3 mM hexanesulfonic acid (pH 3)	Electro-chemical	Analysis of histamine in isolated mast cells.	[215]

Table 6 (continued)

HPLC Conditions for the Analysis of Histamine

15 cm × 4.6 mm i.d. of Asahipak ODP-50 at 40°C	Acetonitrile-50 mM-borate buffer (pH 9.9) (23:77) containing 2 mM-o-phthaldehyde-N-acetyl-L-cysteine (0.5 mL/min)	Fluorimetry at 430 nm (excitation at 330 nm)	Analysis of amines in food	[216]
25 cm × 4.6 mm i.d., (5 μm) ODS basic column, at 60°C.	1% THF in 0.05 M sodium acetate-methanol (9:11 to 1:4 in 25 min)	Fluorimetry at 445 nm (excitation at 330 nm)	Analysis of biogenic amines in wines	[217]
12.5 m × 2 mm i.d. of Superspher 100 RP-18	Methanol-aqueous 7% triethylamine acetate of pH 7.5 (17:3), (6.2 mL/min)	Fluorimetry at 450 nm (excitation at 340 nm)	Identifies biogenic amine-producing bacterial cultures using isocratic HPLC.	[218]
15 cm × 4.6 mm i.d. of Inertsil C4, 5 μm, at 30°C.	5 mM-SDS in 10 mM $HClO_4$-acetonitrile (17:3 at 0 min to 4:1 at 20 min to 13:7 at 50 min, then 17:3 for 30 min) (1-2 mL/min).	UV 210 nm.	Separation and assay of biogenic amine by ion-pair HPLC.	[219]

Table 6 (continued)

HPLC Conditions for the Analysis of Histamine

Column	Mobile phase and flow rate	Detection	Remarks	Ref.
15 μm × 3.9 mm i.d., (4 μm) Nova Pak C-18 with a guard column of the same material.	Sodium acetate-sodium octane sulfonate-acetonitrile (1 mL/min)	Fluorimetry at 445 nm (excitation at 340 nm)	Analysis of biogenic amines in meat and meat products	[220]
15 cm × 4.6 mm i.d. of 3 μm Spherisorb 3S TG	Acetonitrile (65-90%) in water (0.8 mL/min)	UV detection	Analysis of biogenic amines in fish and meat products	[221]
12.5 cm × 2.5 mm i.d. (5 μm) of Lichrospher 100 RP-18	Methanol-water [1:1 (held for 30s.) to 17:3 over 6.5 min (held for 5 min), and to 1:1 over 2 min]	UV at 254 nm	Analysis of biogenic amines in fish implicated in food poisoning.	[222]
25 cm × 4.6 mm i.d. 5 μm Superspher 60 RP-18e	100 mM-sodium acetate buffer at pH 4.4-acetonitrile (1.1 mL/min)	Fluorimetry at 335 nm (excitation at 274 nm)	Analysis is biogenic amines in meat and meat products	[223]

Table 6 (continued)

HPLC Conditions for the Analysis of Histamine

Column	Mobile phase and flow rate	Detection	Remarks	Ref.
25 cm x 4.6 mm i.d., C8 column	0.04 M-KH_2PO_4 – H_3PO_4 buffer of pH 3.5 in acetonitrile (3:7), (1 mL/min)	Fluorimetry at 444 nm (excitation at 350 nm)	Analysis of amines in biological fluids	[224]
15 cm x 4.6 mm i.d. of Shodex Asahipak ODP-50 at 40°C	50 mM borate buffer of pH 10.	Fluorimetry at 430 nm (excitation at 330 nm)	Analysis of histamines in chicken tissues	[225]
25 cm x 4.6 mm i.d. of a Shandon RP-18	0.5 mM ammonium acetate buffer and acetonitrile-0.5 mM-ammonium acetate buffer (4:1)	UV at 271 nm	Measurement of biogenic amines	[226]
25 cm x 4.6 mm i.d. of ODS RP-18 (5 μm)	20-80% Methanol in acetate buffer (pH 6.2), (1.2 mL/min)	–	Analysis of histamine in Monterey Sardine muscles and canned products	[227]

Table 6 (continued)

HPLC Conditions for the Analysis of Histamine

Column	Mobile phase and flow rate	Detection	Remarks	Ref.
25 cm x 4.6 mm i.d. (5 μm) of Spherisorb ODS-2 at 34-35°C	30% Acetonitrile in 20 mM-KH_2PO_4 of pH 4 (0.8 mL/min)	Fluorimetry at 415 nm (excitation at 315 nm)	Analysis of histamine in tuna fish samples.	[228]
15 cm x 4.6 mm i.d. of a Cosmosil 55L	Chloroform-DMF-water (210:90:4) containing 0.4% acetic acid (0.8 mL/min)	Colorimetric at 423 nm.	Used to assay histidine decarboxylase inhibitory activity of soy flavones from soy sauce	[229]
25 cm x 4.6 mm i.d. (5 μm) of Spherisorb S5 ODS 2 with a guard column (1 cm × 4.6 mm i.d.)	9 mM Phosphate buffer containing 8.3 mM heptane-sulfonate (A) methanol (B) [from 100% A (held for 1 min) to 74% in 3.5 min, to 65% in 6 min, to 85% (held for 15.5 min) in 1 min and back to 100% A (held for 10 min)]	UV at 215 nm	Used for the analysis of cheese wine, beer, tofu and soy sauce.	[230]

Table 6 (continued)

HPLC Conditions for the Analysis of Histamine

Column	Mobile phase and flow rate	Detection	Remarks	Ref.
25 cm x 4.6 mm i.d. (10 μm) of Partisil ODS	Sodium acetate buffer of pH 4.9-acetonitrile (7:3), (1 mL/min)	Electro-chemical	Analysis of histamine in human plasma	[231]
25 m × 4 mm i.d. of GROM-SIL Polyamin-2	Acetonitrile-water-acetic acid (100:400:3) solvent A and acetonitrile-water-acetic acid (475:25:3) solvent B	Fluorimetry at 310 nm (excitation at 265 nm)	Analysis of histamine in cheese	[232]
25 cm × 4.5 mm i.d. of TSK gel ODS-80	Methanol-water (10:3) (1 mL/min)	Fluorimetry at 470 nm (excitation at 415 nm)	Histamine and other amines.	[233]

An automated ion exchange method was used for the composite analysis of histamine and biogenic amines in cheese [238]. The method was reported to be simple and accurate, and was used to assess the efficiencies of extraction of the amine from mild or old cheddar cheese, by trichloroacetic acid, $HClO_4$, and methanol. Histamine was detected in the old cheese.

Histamine and other biogenic amines were determined in wine with an amino acid analyzer [239]. Wine was concentrated, mixed with sodium citrate buffer (pH 1.8), filtered, and centrifuged. The supernatant solution was analyzed on an amino acid analyzer with use of the ninhydrin reaction. The exchanger column was packed with Durrum DC 6A resin, and operated at 62.5°C for 35 minutes, and then at 79.5°C for 38 minutes with two 0.2 M sodium citrate-sodium chloride buffers containing methanol or ethanol. Histamine was determined with a recovery of 96 to 100%.

Simon-Sarkadi and Holazapfel determined histamine in leafy vegetables using an amino acid analyzer [240]. Homogenized vegetable tissue was extracted with 10% trichloroacetic acid (2 × 20 mL) at 5°C. The extracts were centrifuged, combined, adjusted to pH 5.5 with KOH, and diluted to 50 mL. After filtration, portions (0.25 mL) were analyzed on a column of BTC 3118 cation-exchange resin (K^+ form) at 65°C, with a mobile phase (0.75 mL/min) of citrate buffer, and post-column reaction with ninhydrin for detection at 570 nm.

Trevino *et al* reported the determination of histamine in mini-salami during long-term storage [241]. Sausage mixture (5 gm) was homogenized with 45 gm of 10% trichloroacetic acid, the mixture was filtered, and the filtrate (4 mL) mixed with 50 mL of 0.1 M sodium acetate adjusted to pH 6 with acetic acid (buffer A). The resulting solution was readjusted to pH 6 with NaOH and applied to a column packed with Amberlite CG 50 cation exchanger. After washing with 50 mL of buffer A, the analytes were eluted with 20 mL of 0.2 M HCl. A portion (1 mL) of the eluent was analyzed at 62-79°C on a Durrum DC 6a column (4 cm × 9 mm i.d.) with various buffers of pH 5.4-6.8 (containing sodium citrate, NaCl, Brij 35, phenol, methanol, and ethanol) as mobile phases (45 mL/h). Detection was at 570 nm after online post-column derivatization with ninhydrin. The method was used to study the formation of biogenic amines in raw meat products during the maturity process and storage.

4.9.8. Gas Chromatography-Mass Spectrometry

The determination of N-methylhistamine in tissues by gas chromatography-mass spectrometry and the GC-MS characteristics have been reported [242]. Mita *et al* determined histamine in biological materials by GC-MS [243]. Simultaneous analysis of histamine and derivatives in human plasma and urine by GC-MS, have been described [244]. The authors have also reported a quantitative determination of histamine and N-methyl histamine in human plasma and urine by GC-MS [245].

Hough *et al* reported an improved GC-MS method to measure N. tau-methylhistamine [246]. Capillary gas chromatographic-mass spectrometric determination of histamine, in tuna fish causing scombroid poisoning, has been published [247]. Keyzer *et al* determined N. tau. methylhistamine in plasma and in urine by isotope-dilution-mass spectrometry [248]. The identification and fragmentography determination of N tau. methyl histamine, in cerebrospinal fluid, by GC-MS, has been reported [249]. The authors [250] have also identified and determined 1-methylimidazol-4-yl-acetic acid (a histamine metabolite), in human cerebrospinal fluid by GC-MS.

Roberts and Oates have described an accurate and efficient method for the quantification of urinary histamine by gas chromatography-negative ion chemical ionization mass spectrometry [251]. Histamine was extracted from urine, to which $[^2H_4,]$-histamine had been added as an internal standard, into butanol-heptane, then back-extracted into HCl and derivatized with a-bromopentafluorotoluene. The derivative was extracted into dichloromethane and determined by GLC on a 2 foot column packed with Poly I-110 and operated at 250°C. Methane was used as the reagent and carrier gas, and 250 eV negative-ion chemical ionization mass spectrometry was carried out while monitoring the ratio of the ion intensities at m/e 430 and 434.

Roberts reported the use of the pentafluorobenzyl derivative of histamine for the determination of histamine in the biological fluids by gas chromatography - negative ion channel ionization mass spectrometry [252].

Keyzer *et al* reported the methodology and normal values for the measurement of plasma histamine by stable-isotope dilution gas chromatography-mass spectrometry [253]. Plasma containing (^{15}N)-histamine as the internal standard, and buffered at pH 9, was cleaned up on a Sep Pak silica column. The HCl-methanol eluent was evaporated to dryness, the residue was dissolved in sodium hydroxide solution, and this solution extracted with chloroformbutanol (4:1). The extract was evaporated to dryness and the residue treated with acetonitrile heptafluorobutyric anhydride. The resulting derivatives were analyzed by GC-MS on a fused silica column (12.5 m x 0.22 mm) coated with SP-2100 and deactivated with Carbowax. The temperature was programmed from 150 to 220°C at 10°C/min, and helium was used as carrier gas (0.5 mL/min). The mass spectrum was carried out with 120 eV chemical ionization (NH$_3$ as reagent gas), and the ions at m/e 366 and 368 were monitored.

The comparative evaluation of a radioenzymic method for the determination of urinary histamine with a mass spectrometric method has been reported [254]. Interference from urinary substances was observed for the radioenzymic, and up to a 34-fold difference was found between the results of the two methods. The GC-MS method yielded the higher results.

Keyzer *et al* reported the measurement of N-tau. methylhistamine concentrations in plasma and urine as a parameter for histamine release during anaphylactoid reactions [255]. Histamine and N tau. methylhistamine were separated from plasma and urine by chromatography on a Sep Pak silica column. N tau. methylhistamine was extracted into chloroform, and histamine was subsequently extracted into chloroform-butanol from a sodium chloride saturated alkaline solution. After evaporation of the solvents, histamine and derivatives were reacted to N-α-heptafluorobutyryl-N-tau-methoxy carbonyl histamine and N-tau.-methyl-N-α-, 2-bis (pentafluoropropionyl) histamine, respectively. The derivatives were cleaned up, and their ethyl acetate solution was co-injected into a Varian 3700 gas chromatograph coupled to a Finnigan MAT 44-S quadrupole mass spectrometer operated in the NH$_3$ chemical ionization mode. Histamine and its internal standard [(^{15}N$_2$)-histamine] were monitored at m/e 366 and 368, respectively.

The analysis of histamine and N. tau.-methylhistamine in plasma by gas chromatography-negative-ion chemical ionization mass spectrometry was described [256]. The ^{2}H-labeled analogues were added to plasma as internal standards. For histamine, plasma was deproteinized and histamine extracted into butanol, and then back extracted into HCl before conversion into its pentafluorobenzyl derivative. Cleanup was effected on a silica gel column. The GC separation was carried out on a column (10 cm × 0.2 mm) of DB 1 with helium as the carrier gas (50 cm/s), with temperature programming from 100°C at 4°C/min to 140°C, and then at 35°C/min to 300°C. Negative ion CI-MS with selected-ion monitoring at m/e 450 and 434 was used for detection.

Murray *et al* described an assay method for N tau. methylimidazoleacetic acid (1-methylimidazol-4-yl-acetic acid), a major metabolite of histamine, in urine and in plasma using capillary column gas chromatography-negative ion mass spectrometry [257]. Another gas chromatographic-mass spectrometric assay, to measure urinary N-tau. methylhistamine excretion in man, has been reported [258]. Urine was treated with deionized water (adjusted to pH 8 with ammonia) and N. tau. $[^2H_3]$-methylhistamine (internal standard) and was passed through a column of Bond Elut CBA. The compound and the internal standard were eluted with 0.1 M HCl. After evaporation to dryness under nitrogen, the residue was derivatized with 3,5-bis (trifluoro-methyl-benzoyl) chloride. A portion of the derivatized solution were analyzed on a column (30 m x 0.25 mm) of DB5, temperature programmed from 200°C (held for 1 minute) to 320°C at 20°C/minute with helium as the carrier gas. Negative ion mass spectra at m/e 605 and 608 were monitored for the compound and internal standard, respectively, using ammonia as the reagent gas.

4.9.9. Isotachophoresis

Histamine was determined in stinging insect venoms by isotachophoresis [259]. Histamine has been assayed in fish and in canned fish by capillary isotachophoresis [260]. Measurement of histamine in stinging venoms by isotachophoresis has also been reported [261].

Histamine was determined in biological fluids by capillary isotachophoresis, and by fluorescence [262]. An isotachophoresis analyzer was equipped with a potential gradient detector and a UV detector. Analysis was carried out in a PTFE capillary (20 cm × 0.5 mm). The leading electrolyte comprised 5 mM KOH, adjusted to pH 7 with morpholino-ethanesulfonic acid. The terminating electrolyte was 5 mM creatinine, adjusted to pH 4.6 with 1 M HCl. To increase the sensitivity of the method for the determination of histamine in serum, the analyzer was modified by including a spectrofluorimeter, so that the capillary replaced the cell of the spectrofluorimeter. An application valve replaced the micro syringe injection port. Histamine was extracted from serum samples with butanol, and condensed with phthalaldehyde. Levels as low as 3 ng/mL of histamine in serum could be determined.

4.9.10. Electrophoresis

Priebe reported the use of a thin layer electrophoresis method for the determination of histamine in fish and in fish products [263]. The significance of use of histamine and amino acids for the elution of non-histone proteins in copper chelate chromatography has been published [264]. Kamel and Maksoud described micro-electrophoretic and chromatofocusing techniques for quantitative separation and identification of histamine and other imidazole derivatives [265].

The synthesis of a new acrylamido buffer (acryloylhistamine) for isoelectric focusing in immobilized pH gradients and its analysis by capillary zone electrophoresis, was reported [266]. Nann et al [267] developed a quantitative analysis in capillary zone electrophoresis (CZE), using ion-selective microelectrode (ISE) as on-column detectors for the determination of histamine. Altria et al reported a capillary electrophoresis method for the assay of histamine acid phosphate [268]. Aqueous solutions containing histamine were analyzed by capillary electrophoresis using a Beckman P/ACE 2200 or 5100 system with capillaries (37 cm × 75 μm i.d.) operated at 30°C and 13 kV using 25 mM-NaH_2PO_4 adjusted to pH 2.3 with H_3PO_4 as electrolyte and detection at 200 or 214 nm.

4.9.11. Radioenzymic Assay

Bruce et al reported an improved radioenzymic assay for determination of histamine in human plasma, whole blood, urine, and gastric juice [269]. Application of thin layer chromatography to histamine radioenzymic assay in plasma was reported [270]. Guilloux et al described an enzymic isotopic method for the assay for histamine [271]. Verburg et al determined histamine by a new radio enzymic assay procedure using purified histamine N-methyltransferase [272].

Warren et al measured the urinary histamine using both fluorimetric and radio isotopic enzymic assay procedures [273]. Histamine N-methyl transferase and [^{3}H methyl] or [^{14}C methyl]-s-adinosyl-methionine as the substrate. The enzyme was extracted from rat kidney [274] and purified by ammonium sulfate precipitation and dialysis. Although the calibration graph of activity against concentration was linear, the reagent blank produced counts of about 60% of the standard without any added histamine. The assay was dependent on enzyme concentration and as the activity was still increased in heat-denatured crude enzyme preparations, the activity was due to the endogenous histamine.

Brown *et al* described a radioenzymic assay for measurement of tissue concentrations of histamine [275]. Histamine was tested for its adherence to a mechanical homogenizer commonly used in extraction of tissue samples. During homogenization for 60 seconds, 15 to 17% of the histamine originally present could not be recovered, owing to reversible binding to the homogenizer. The initial addition of [^{14}C]-histamine to the sample and measurement of the disappearance of radioactivity during homogenization permitted correction for binding to the homogenizer. This correction technique was validated by the measurement of endogenous concentration of histamine in the tracheal posterior membranes of dogs.

Simultaneous analysis of histamine and N-α-methylhistamine in biological samples by an improved enzymic single isotope assay has been described [276]. Chevrier *et al* used an enzyme immunoassay procedure for the measurement of histamine in blood and in biological tissues [277]. Histamine was converted into its 1,4-benzoquinone addition product, which competes with histamine-peroxidase conjugate for monoclonal antibody bound to a microtitre plate. Peroxidase was assayed with o-phenylenediamine-hydrogen peroxide. Levels as low as 300 pg/mL of histamine could be determined in blood or in tissue homogenates. An enzymo-immunometric assay was also described, involving the use of plates coated with histamine-ovalbumin conjugate and peroxidase-labeled sheep anti-mouse IgG [277].

Rauls *et al* reported a modification of the enzyme isotope assay [278]. Plasma was incubated in an ice bath for 90 minutes with [^{3}H]-S-adinosylmethionine, histamine methyltransferase, and water. After incubation with acetonitrile containing 1-methylhistamine and centrifugation, the supernatant solution was subjected to thin layer chromatography on silica gel with acetone-aqueous ammonia (23:2) as the mobile phase, and detection by treatment with iodoplatinate spray. 1-

methyl histamine was determined by liquid scintillation counting. The calibration graph was linear for 0.1 to 10 ng/mL of histamine.

The preparation, characterization of monoclonal antihistamine antibody, and application to enzyme immunoassay of histamine, was reported [279]. Antibodies of histamine were raised in mice against a histamine bovine serum albumin conjugate prepared with 1,4-benzoquinone as the coupling agent. The characterization of the antiserum was described. To determine histamine, microtitre plates were coated with histamine-bovine serum albumin conjugate, and incubated with the antibodies of histamine. After removal of unbound reagents, β-D-galactosidase antimouse IgG (raised in sheep) conjugate was added, and the enzyme activity measured with the use of o-nitrophenyl-galacto-pyranoside as substrate and spectrophotometric monitoring at 414 nm.

Optimization of the histamine radioenzyme assay with purified histamine methyltrasferase has been published [280]. The optimal conditions for the radioenzymic analysis of histamine were incubation (at 20°C for 90 minutes at pH 8.3) of a reaction mixture containing 20 μM-[methyl-^{3}H]-S-adenosyl-L-methionine buffer solution of pH 8.3, 40 mM dithiothreitol, 40 mM K_2EDTA of histamine methyltransferase (containing 64 μg/mL of protein), sample (plasma, blood, skin biopsy) solution and water. The reaction was stopped by the addition of 2.5 M $HClO_4$, and the mixture was centrifuged. The supernatant solution was shaken for 5 minutes with 10 M NaOH and tolueneisoamyl alcohol (4: 1). After centrifugation, the organic phase was transferred to a scintillation phial containing ACS solution, and the radioactivity was counted with LKB 1215 Rackbeta liquid scintillation counter with automatic quench correction.

Morel *et al* described an immunoenzyme procedure for the assay of histamine [281]. Histamine was acylated with use of activated succinyl glycinamide at pH 3, and the derivative was incubated overnight at 4°C in microtitre plate wells coated with monoclonal antibodies raised against histamine succinyl glycyl albumin and histamine succinyl glycyl albumin-acetylchloinesterase as a tracer. The plates were washed and bound tracer was detetermined by incubating the plates for 30 minutes at room temperature with substrate containing 0.75 mM acetylthiocholine iodide, 0.5 mM dithio-bis-(2-nitrobenzoic acid), and 30 mM sodium chloride in phosphate buffer (pH 7.4). The reaction was stopped by the addition of tacrine, and the absorbance measured at 410 nm. The enzyme immunoassay method was found to be equally sensitive as a radioimmunoassay, and results obtained by the two methods correlated well (r = 0.93).

An enzyme immunoassay method for the determination of histamine in beer has been reported [282]. Beer was mixed with $HClO_4$ and centrifuged, and the supernatant solution analyzed by enzyme inimunoassay in microtitre wells coated with monoclonal antibodies raised against histamine-benzoquinone. After incubation at room temperature with histamine-benzoquinone-peroxidase, peroxidase activity was measured.

Rauch et al analyzed histamine in foods, by an enzyme immunoassay method [283]. Histamine, present in food samples, was treated with 1,4-benzoquinone to form histamine benzoquinone, by a simple and quick reaction, and a histamine benzoquinone horseradish peroxidase conjugate was used as the labeled hapten. The apparent association constant of the antibody used was 3.6×10^6 P/mol, and Gibbs' energy of the immune complex formation was estimated to find the optimal incubation time of the assay. The method enabled determination of histamine in fish, cheese, wine, and beer at a concentration as low as 7 ng/mL with an accuracy of ± 15%. Hegstrand reported a direct and sensitive microassay method for the analysis of the mammalian histidine decarboxylase [284].

4.9.12.   Radioimmunoassay

The interference of iodine-125 ligands in the radioimmunoassay (RIA) of histamine in serum, and an evidence implicating thyroxine binding globulin, was reported [285]. A radioimmunoassay method for the determination of histamine in plasma was also reported [286]. Antibodies were raised in rabbit against histamine-protein conjugates, produced by using 1,6-diisocyanatohexane as coupling agent. The conjugates of haemoglobin or bovine or human serum albumin were suitable for use in the RIA of histamine in biological fluids. Plasma was filtered through an Amicon membrane in a Centricon micro concentrator. The filtrate was mixed with N-α-acetyl-L-lysine-N-methylamide in phosphate buffer solution (pH 10.5), and 1,6-diisocyanatohexane was added. The mixture was centrifuged and the supernatant solution analyzed by RIA with use of ^{125}I-labeled histamine, 1,6-diisocyanatohexane glycyltyrosine as the radio ligand, and polystyrene spheres coated with antiserum.

Nambu *et al* reported that Ozagrel (a thromboxane A_2 synthetase inhibitor) inhibits anaphylactic bronchoconistriction and reduces histamine level in bronchoalveolar lavage fluid in sensitized guinea pigs [287]. Labeled ^{3}H-thromboxane and 6 oxo $PGF1\alpha$ (50 μL) (each 400 dpm), water (1 mL), and 0.1 M acetic acid (1 mL) were mixed with plasma (1 mL), and the samples centrifuged at 1700 G for 5 minutes. The supernatant liquid was extracted on a Sep Pak C18 column, followed by a Bond Elut DEA column, and finally with the Sep Pak column again. The products were separated by TLC on silica gel, using a solvent system of ethyl acetate / isooctane-acetic acid / water (9:5:2:10). The products were visualized with 10 to 20% phosphomolybdic acid, and ^{3}H thromboxane and 6 oxo $PGF1\alpha$ were eluted with methanol-acetic acid (99.5:0.5). ^{3}H thromboxane and 6 oxo $PGF1\alpha$ were determined by RIA with dextran-coated charcoal separation after the reaction. The histamine levels were measured by a histamine RIA kit (immunotech). Oosting and Keyzer reported the measurement of urinary N methyl histamine excretion, and have correlated the radioimmunoassay with gas chromatography-mass spectrometry [288].

A comparison of available radioimmunoassay methodologies for the measurement of human lung histamine has been made [289]. Histamine was determined in bronchoalveolar lavage fluid and lung extracts by the Immunotech histamine radioimmunoassay kit according to the manufacturer's instructions. The results correlated well (r = 0.86) with those obtained by a single isotope enzyme assay for histamine in bronchoalveolar lavage fluid, and by spectrophotometric determination of histamine in the lung.

Hammar *et al* described an immunoassay method for the assay of histamine based on monoclonal antibodies [290]. Standard or sample solutions (100 μL) were mixed with 100 μL each of each of ELISA tracer solution (β-galactosidase-1-[N-[3-(2-pyridyldithio)propionyl]-2-aminoethyl}-4-(2-aminoethyl)imidazole conjugate, or X) or radioimmunoassay tracer (^{125}I-labeled histamine-X-conjugate) and monoclonal antihistamine antiserum. After incubation for 18 hours at room temperature, free and bound tracer were separated with anti-mouse IgG-agarose beads, and the mixtures were shaken for 1 hour. For ELISA, *o*-nitrophenyl-β-galactoside solution (200 μL) was added to the pellet, and the mixture incubated for 2.5 hours at 37°C before the reaction was stopped by the addition of 0.5M sodium carbonate (1 mL). After centrifugation, the absorption of the supernatant solution was measured at 420 nm. For radioimmunoassay, the radioactivity of the pellet obtained after treatment with the anti-mouse IgG conjugate was counted in a gamma counter. Brown *et al* reported a sensitive and specific method for measurement of plasma histamine in normal individuals [291].

Serrar *et al* reported the development of a monoclonal antibody-based ELISA method for the determination of histamine in food [292]. The method was applied to the fishery products, and was compared with the HPLC assay. The homogenized fish (1 g) was extracted with trichloroacetic acid (1 mL). The extracts were derivatized with 1,4-benzo-quinone in ethanol (3 mg/100 μL) and phosphate buffer (pH 6), and then incubated at 20°C for one hour. The pH was adjusted to pH 7.4 with 100 μL of 3.3 M-triethanolamine containing 0.7 M-glycine. The solution was diluted (1 in 10) with PBS/Tween. Four antihistamine mAbs were raised, and purified by standard ion-exchange chromatography and gel filtration. Each purified mAbs (100 μL) was mixed with 100 μL of the prepared sample and incubated at 20°C for 1 hour. A portion (100 μL) of each mixture was transferred to a microtiter plate well that was coated with histamine-benzoquinone-casein conjugate and incubated at 20°C for 30 minutes. Unbound proteins were removed with PBS/Tween, and the immobilized mAb was detected with peroxidase-labeled anti-mouse IgG antibody. Calibration graphs obtained were linear from 10-100 ng/mL of histamine.

Krueger *et al* determined histamine in fish by ELISA [293], and the results of this study agreed well with those obtained by isotope dilution GC-MS and HPLC. The comminuted sample was homogenized with 10% $HClO_4$, and the filtered homogenate was diluted with water. The solution was treated with 3-(4-hydroxyphenyl)-propionic acid N-hydroxysuccinimide at 37°C for 30 minutes, and the mixture diluted with assay buffer before being transferred to an avidin-coated microtitre plate. The processing was completed by the addition of biotinylated histamine anti-N-acylhistamine antiserum raised in the rabbits. After incubation overnight at 4°C, the plate was washed before adding an alkaline phosphate-goat anti-rabbit antiserum conjugate, and, after one hour at room temperature, assay of bound enzyme with 4-nitrophenyl phosphate and absorbance measurement at 405 nm.

Weise *et al* compared the ELISA, electrophoresis, and HPLC methods for the determination of histamine contents of some selected foods [294]. The histamine contents of tuna in oil, anchovy fillets, sauerkraut, and red wine were determined by ELISA, HPLC, and thin layer electrophoresis.

Agreement of results was best between ELISA and HPLC. Electrophoresis and ELISA were recommended for use in the screening of small and large numbers of samples prior to the determination of histamine in positive samples by a modified form of the HPLC method. The modifications include the adjustment of pH of the total extract to 5.5 with acetate buffer. A UV detector should be used for histamine detection. The methods were applied to a reference histamine-contaminated anchovy fillet. The results of the HPLC agreed with the given values, while those of ELISA and electrophoresis were higher and lower, respectively, than those of the certified values.

Aygun et al also reported a comparison of ELISA and HPLC for the determination of histamine in cheese [232]. Cheese was homogenized with PBS, centrifuged, filtered, and the supernatant diluted with PBS for a competitive direct ELISA with absorbance measurement at 450 nm.

4.9.13.   Miscellaneous Methods

Histamine was measured in plasma, and an improved method and normal values were described [295]. Histamine was determined, in casein, tuna, serum, and urine using a single column amino acid analyzer [296]. The drug was measured in plasma, in a quality control study [297]. The effect of adsorption of histamine to glass surface on its estimation was also reported [298].

5.   Stability

Marwaha et al developed and evaluated a quantitative, colorimetric, simple stability-indicating assay method for histamine solutions [299]. The method used the Pauly reaction, which is specific for the imidazole group. Stock solution of histamine diphosphate were diluted with isotonic phosphate buffer to produce standards for an imidazole-group assay, and of the free amino group for the USP assay method that is based on the Folin reaction. The specificity of the assay was evaluated by subjecting samples to ultraviolet irradiation and heat for various time periods, and analyzing for histamine content. Both methods were found to provide reproducible linear calibration curves passing the origin.

Pratter *et al* studied the stability of histamine diphosphate solutions as a function of time, concentration, fluorescent, light exposure, and sterility [300]. The study found that histamine diphosphate solutions (2.76 and 22.1 mg/mL, or, 1 to 8 mg/mL in histamine base) showed no evidence of degradation over a four-months period when kept at 12°C, unless gas sterilization techniques were used in preparing the solutions. Bacterial contamination was frequent, and at lower concentrations, histamine solutions became contaminated with bacteria, and showed complete degradation within 9 to 11 weeks of their preparation.

The long-term stability of histamine in sterile broncho-provocation solutions stored under different conditions was studied by Marwaha and Johnson [301]. Histamine solutions (2 and 8 mg/mL) were stored in specially adopted translucent and black plastic bottles, and kept in a cool place (12°C) and exposed to fluorescent light. Multiple aliquots of each concentration were also stored at -20°C in sterile polypropylene tubes with snap caps, and were covered with aluminum foil. The contents of histamine in the stored samples were determined by a stability indicating colorimetric assay. Solutions were found to be stable for up to eight months at 12°C when stored in either translucent or black plastic bottles.

Nielsen *et al* studied the stability of histamine dihydrochloride in the solution phase [302]. The activity associated with bacterial or fungal contamination of the stored histamine dehydrochloride dilutions were tested after storage at 20°C, 4°C, and -18°C. The dilutions with a concentration of and below 0.25 mg/mL had a significantly reduced activity after one month of storage at 20°C. The activity of the dilutions stored at either 4°C or -18°C were stable for at least 6 months.

McDonald *et al* reported that histamine acid phosphate solutions, which are usually used in tests for asthma and a positive control, in general allergenic testing, are stable after sterilization by heating in an autoclave [303]. Authors have shown that histamine acid phosphate solutions can be sterilized successfully by heating in an autoclave with little degradation, and that subsequent storage of autoclaved solutions indicates a minimum shelf-life of four months.

The degradation of histamine solutions that are used for broncho-provocation was studied by Marshik *et al* [304]. Authors have concluded that exposure to fluorescent light at room temperature results in degradation of histamine solution used for broncho-provocation.

Dilutions stored in unit dose syringes and protected from light are stable for at least eight weeks in the refrigerator, and up to 12 months frozen. Once removed from the refrigerator or freezer, the solutions should be used within six hours or discarded.

6.   Pharmacology, Drug Metabolism, and Pharmacokinetics

6.1   History

β-aminoethylimidazole (histamine) was first detected as a uterine stimulant in ergot extracts, and proven to be derived from the bacterial action on ergot. Histamine was found to stimulate smooth muscles, and to possess an intense vasodepressor action. Histamine acts through more than one receptor, and at least three distinct classes of receptors have been identified (designated as H_1, H_2 and H_3). The H_1 receptor is blocked selectively by the classical antihistamines, while the H_2 receptor has been found to control gastric secretions. The H_3 receptor appears to exist only in the CNS [12-15].

6.2   Distribution

Histamine is widely distributed throughout the animal kingdom, and is present in venoms, bacteria, and plants [16]. Almost all mammalian tissues contain histamine in amounts ranging from less than 1 μg/g to more than 100 μg/g. Concentrations in plasma and other body fluids are generally very low, but human cerebrospinal fluid contains significant amounts [17]. The mast cell is the predominant storage site for histamine in most tissues, so the concentration of histamine is particularly high in tissues that contain large numbers of mast cells (skin, the mucosa of the bronchial tree, and the intestinal mucosa) [18].

6.3   Biosynthesis, Storage and Degradation

Every mammalian tissue that contains histamine is capable of synthesizing the substance from histidine by virtue of their (L)-histidine decarboxylase content. As shown in Figure 11, there are two major paths of histamine metabolism in man. The more important of these involves ring methylation, and is catalyzed by the enzyme histamine-N-methyltransferase. Most of the product (N-methylhistamine) is converted by monoamine oxidase (MAO) to N-methyl imidazole acetic acid. The products are imidazole acetic acid and eventually its riboside, with the metabolites being excreted in the urine [18].

Figure 11. Major metabolic pathways of histamine. **HMT**: histamine N-methyltransferase; **MAO-B**: monoamine oxidase type B; **DAO**: diamine oxidase; **ALDH**: aldehyde dehydrogenase; **ALO**: aldehyde oxidase; **XO**: xanthine oxidase; **PRT**: phosphoribosyltransferase [18].

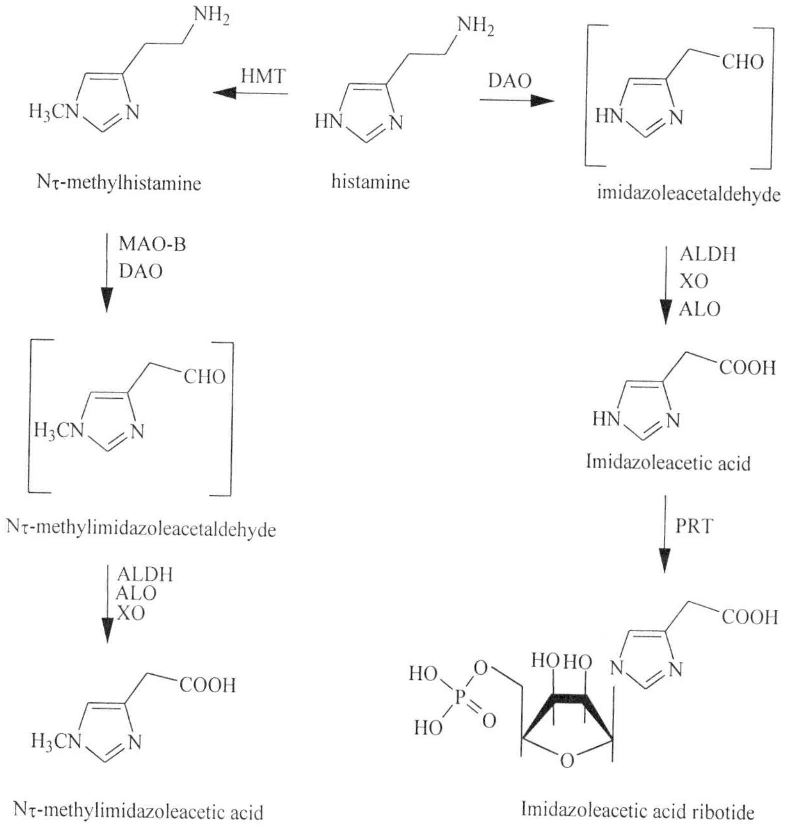

6.4   Pharmacological Effects

Once released, histamine can exert local or widespread effects on smooth muscles and glands. Histamine contracts many smooth muscles (such as, those of the bronchi and gut), but powerfully relaxes others (including small blood vessels). It is also a potent stimulus to gastric acid secretion. There are other effects, such as formation of edema and stimulation of sensory nerve endings. Many of these effects, such as bronchoconstriction and contraction of the gut, are mediated by H_1 receptors [13], which are readily blocked by pyrilamine and other classical antihistamines. Other effects, most notably gastric secretion, are the results of activation of H_2 receptors, and can be inhibited by H_2 antagonists [15]. Some responses, such as hypotension that results from vascular dilatation, are mediated by both H_1 and H_2 receptors. H_3 receptor appears to exist only in the CNS [18].

Acknowledgement

The authors wish to thank Mr. Tanvir A. Butt, Department of Pharmaceutical Chemistry, College of Pharmacy, King Saud University, Riyadh, Saudi Arabia, for the typing of this profile.

References

1.   *The Merck Index* (12th edn), S. Budavari, ed, Merck & Co., Inc., Whitehouse Station, N.J., 1996, No. 8756, p. 807.

2.   *Clarke's Isolation and Identification of Drugs* (2nd edn), A.C. Moffat (ed.), The Pharmaceutical Press, London, 1986, p. 659. (a) *ibid*, 1st edition, 1978, volume 1, page 364.

3.   *Martindale, The Extra Pharmacopoeia* (31st edn), Royal Pharmaceutical Society, London, 1996, p. 1103.

4.   A. Gringauz, *Introduction to Medicinal Chemistry*, Wiley-VCH, Inc., N.Y., 1997, p. 621.

5.   T. Nogrady, *Medicinal Chemistry, a Biochemical Approach*, 2nd ed., Oxford University Press, Inc. N.Y., 1988, p. 211.

6.   G.L. Patrick, *An Introduction to Medicinal Chemistry*, Oxford University Press, Inc., N.Y., 1995, p. 284.

7.   Pyman, *J. Chem. Soc.*, **99**, 668 (1911), through reference 1.

8.   Koessler and Hanke, *J. Amer. Chem. Soc.*, **40**, 1716 (1918), through reference 1.

9.   Garforth and Pyman, *J. Chem. Soc.*, **489**, (1935), through reference 1.

10.   Levy-Bruhl and Ungar, *Ann. Inst. Pasteur*, **61**, 828 (1938), through reference 1.

11.   T.B. Paiva, M. Tominaga and A.C.M. Pavia, *J. Med. Chem.*, **13**, 689 (1970).

12.   H. Dale and P. Laidlaw, *J. Physiol.*, **49,** 318 (1910).

13.   A. Ash and H. Schild, *Br. J. Pharmacol.* **27,** 427 (1966).

14.   J. Black, W. Duncan, C. Durant, C. Ganellin, and E. Parsens, *Nature* **236**, 385 (1972).

15.   J. Arrang, M. Garbarg, J. Lancelot, J. Lecomte, H. Pollard, M.
       Robba, W. Schunack, and J. Schwartz, *Nature* **327**, 117 (1987).

16.   O. Reite, *Physiol. Rev.*, **52**, 778 (1972).

17.   J. Khandelwal, L. Hough and J. Green, *Klin. Wochenschr.* **60**, 914
       (1982).

18.   Goodman and Gilman's, ***The Pharmacological Basis of
       Therapeutics***, 9th edition, Pergamon Press, Inc., USA, 1996,
       Chapter 25, page 581.

19.   S. Schwedt, *Labor Praxis,* **6**, 1198, 1201, 1204 (1982).

20.   M.A. Beaven, A. Robinson-White, N.B. Roderick and G.L.
       Kauffman, *Klin. Wochenschr.,* **60**, 873 (1982).

21.   J.J. Keyzer, B.G. Wolthers and W. Van-der-slik, *Tijdschr. Ned.
       Ver. Klin.* Chem., **11**, 146 (1986).

22.   T. Seki, *Kagaku, Zokan* **117**, 85 (1990).

23.   W.J. Hurst, *J. Liq. Chromatogr.,* **13**,1 (1990).

24.   W. Lorenz and E. Neugebauer, *Hanb. Exp. Pharmacol.,* **97**, 9
       (1991).

25.   ***The United States Pharmacopeia 23***, United State Pharmacopeial
       Convension, Inc., Rockville, MD, 1995, page 743.

26.   ***The British Pharmacopoeia 1993***, Her Majesty'Stationary Office,
       London, U.K., 1993, Volume 1, page 325.

27.   M. Kudoh, M. Kataoka and T. Kambara, *Bunseki Kagaku,* **28**, 705
       (1979).

28.   I. Karube, I.. Satoh, Y. Araki, S. Suzuki and H. Yamada, *Enzyme
       Microb. Technol.,* **2**, 117 (1980).

29.   B. Yu, *Biosensors,* **4**, 373 (1989).

30.   T. Tatsuma and T. Watanabe, *Anal. Chem.,* **64**, 143 (1992).

31. L. Shi, T.P. Lu, B. Zhou, H.X. Yin and G.H. Yin, *Fenxi Huaxue,* **21**, 1466 (1993).

32. A.M.Y. Jaber, *Anal. Lett.,* **19**, 2039 (1986).

33. M.C. Cago-Agrafojo and J.L. Hidalgo-de Cisneros, *An. Ouim.,* **86**, 305 (1990).

34. J.L. Hidalgo-de Cisneros, A. Marchena Gonzalez, N. Moreno Dias de La Riva and J.A. Munoz Leyva, *Bull. Electrochem.,* **6**, 886 (1990).

35. E.P. Medyantseva, G.K. Budnikov, T.A. Balakaeva and I.A. Zakharova, *Zh. Anal. Khim.,* **46**, 1573 (1991).

36. T.G. Wu, and J.L. Wong, *Anal. Chem. Acta.,* **246**, 301 (1991).

37. R.E. Fornes, R.D. Gilbert, and M.C. Battigelli, *J. Environ. Sci. Health,* Part A, **16**, 289 (1981).

38. Y.L. Xie, J.H. Wang, H. Xie and R.Q. Yu, *Fenxi Huaxue*, **20**, 1023 (1992).

39. T. Sakai, N. Ohno T. Wakisaka and Y. Kidani, *Bunseki. Kagaku,* **31**, 356 (1982).

40. T. Sakai, N. Ohno, T. Wakisaka and Y. Kidani, *Bull. Chem. Soc. Jpn.,* **55**, 3464 (1982).

41. J. Stockemer and M. Stede, *Arch. Lebensmittelhyg.,* **30**, 59 (1979).

42. O.H. Wilhelms, *J. Immunol. Methods,* **36**, 221 (1980).

43. B. Fischer and W. Schmutzler, *Allergologie,* **3**, 356 (1980).

44. Y. Endo, *J. Chromatogr.,* **205**, 155 (1981).

45. G. Myers, M. Donlon and M. Kaliner, *J. Allergy Clin. Immunol.,* **67**, 305 (1981).

46. S.J. Lewis and M.R. Fennessy, *Agents Actions,* **11**, 228 (1981).

47. J.B. Luten, *J. Food Sci.,* **46**, 958 (1981).

48.   Y. Ogawa, M. Kodama, N. Ito, 0. Kodama, Y. Kato, T. Matsuyama and H. Ezaki, *Hiroshima Diagaku Igaku Zasshi,* **29**, 881 (1981).

49.   L. Sekardi and K.D. Freidberg, *Pharmacology,* **24**, 45 (1982).

50.   E.S.K. Assem and E. K. S. Chong, *Agents Actions,* **12**, 26 (1982).

51.   G. Suhren, W. Heeschen and A. Toole, *Milchwissenschaft,* **37**, 143 (1982).

52.   B. Lebel, *Anal. Biochem.,* **133**, 16 (1983).

53.   C. Buteau, C.L. Duitschaever and G.C. Ashton, *J. Chromatogr.,* **284**, 201 (1984).

54.   S. Ramantanis, C.P. Fassbender and S. Wenzel, *Arch. Lebensmittelhyg.,* **35**, 201 (1984).

55.   M.C. Gutierrez, A. Gomez Hens and M. Valcarcel, *Anal. Chem. Acta,* **185**, 83 (1986).

56.   E. Kownatzki, G. Grueninger and N. Fuhr, *Pharmacology,* **34**, 17 (1987).

57.   M.C. Gutierrez, S. Rubio, A. Gomes Hens and M. Valcarcel, *Talanta,* **34**, 325 (1987).

58.   M.C. Gutierrez, S. Rubio, A. Gomes Hens and M. Valcarcel, *Anal. Chem.,* **59**, 769 (1987).

59.   M.C. Gutierrez, S. Rubio, A. Gomes Hens and M. Valcarcel, *Anal. Chem. Acta,* **193**, 349 (1987).

60.   A. Fonberg Broczek, B. Windyga and J. Kozlowski, *Rocz. Panstw. Zakl. Hig.,* **39**, 226 (1988).

61.   M.C. Vidal Carou, M.L. Izquierdo Pulido, and A. Marine Font, *J. Assoc. Off. Anal. Chem.,* **72**, 412 (1989).

62.   P.D. Siegel, D.M. Lewis, M. Petersen and S.A. Olenchock, *Analyst,* **115**, 1029 (1990).

63.   M.C. Vidal Carou, M.T. Veciana Nogues and A. Marine Font, *J. Assoc. Off. Anal Chem.*, **73**, 565 (1990).

64.   J.M. Hungerford, K.D. Walker, M.M. Wekell, J.E. LaRose and H.R. Throm., *Anal. Chem.*, **62**, 1971 (1990).

65.   J.M. Hungerford and A.A. Arefyev, *Anal. Chem. Acta*, **261**, 351 (1992).

66.   M.E. Diaz Cinco, O. Fraijo, P. Grajeda and J. Lozano Taylor, *J. Food Sci.*, **57**, 355, 365 (1992).

67.   R. Gajewska, Z. Gonawiak and A. Lebiedzinska, *Rocz. Panstw. Zakl. Hig.*, **30**, 47 (1979).

68.   M.I. Yamani and F. Untermann, *Int. J. Food Microbiol.*, **2**, 273 (1985).

69.   K.K. Vishwanath, A.S. Rao and M.V. Sivaramakrishnan, *Indian Drugs*, **24**, 453 (1987).

70.   Y. Yamagami, T. Naito, M. Takayanagi, S. Goto and T. Yashiro, *Chem. Pharm. Bull.*, **35**, 3037 (1987).

71.   T. Sakai, N. Ohno, M. Tanaka and T. Okada, *Analyst,* **109**,1569 (1984).

72.   J.J. Keyzer, H. Breukelman, H. Elzinga, B.J. Koopman, B.G. Wolthers and A.P. Bruins, *Biomed. Mass Spectrom.* **10**, 480 (1983).

73.   J. Eagles and R.A. Edwards, *Biomed. Environ. Mass Spectrom* **17**, 241 (1988).

74.   E. Ya. Matveeva, I.E. Kalinichenko and A.T. Pilipenko, *Zh. Anal. Khim.*, **38**, 710 (1983).

75.   M- Katayama, H. Takeuchi and H. Taniguchi, *Anal. Chem. Acta,* **281**, 111 (1993).

76.   P. Cattaneo, and C. Cantoni, *Ind. Aliment* (Pinerolo, Italy), **17**, 303 (1978).

77.   L. Simon-Sarkadi, A. Kovacs and E. Mincsovics, *J. Planar.
       Chromatogr., Mod-TLC*, **10**, 59 (1997).

78.   A. Kovacs, L. Simon-Sarkadi and E. Mincsovics, *J. Planar.
       Chromatogr., Mod-TLC*, **11**, 43 (1998).

79.   G. Calaresu, *Boll. Chim. Unione Ital Lab. Prov.*, **4**, 262 (1978).

80.   J. Storck, P. Denis and J.P. Pabin, *Ann. Pharni. Fr.*, **37**, 257
       (1979).

81.   S. Pongor, J. Kramer and E. Ungar, *High Resolut. Chromatogr,
       Commun.*, **3**, 93 (1980).

82.   S.E. Hansen, P. Albeck and Bundgaard, *Agents Actions*, **15**, 125
       (1984).

83.   K.D.H. Chin and P.E. Koehler, *J. Food Sci.*, **48**, 1826 (1983).

84.   S. Ramantanis, C.P. Fassbender and S. Wenzel, *Arch.
       Lebensmittelhyg.*, **35**, 80 (1984).

85.   R. Wahl, H.J. Maasch and W. Geissler, *J. Chromalogr.*, **329**, 153
       (1985).

86.   Z. Wu, X. Ren, Y. Wang, L. Gan, X. Wang and J. Guo, *Shengwu
       Huaxue Yu Shengwu Wuli Jinzhan*, **1**, 57 (1987).

87.   E. Pyra and J. Iskierko, *Acta Pol. Pharm.*, **46**, 132 (1989).

88.   T.M. Surgova, M.V. Sidorenko, I.S. Kofman and V.B. Vinnitsky,
       J. Planar Chromatogr. Mod. TLC, **3**, 81 (1990).

89.   J. Sherma, D. Raible and K. Brubaker, *J. Planar Chromatogr.
       Mod. TLC*, **4**, 253 (1991).

90.   N.N. Singh, C. Periera and U.M. Patel, *J. Liq. Chromatogr.*, **16**,
       1845 (1993).

91.   A.R. Shalaby, *Food Chem.*, **49**, 305 (1994).

92.   A.R. Shalaby, *Food Chem.*, **52**, 367 (1995).

93. Y. Wang, S.J. Tan and X.S. Shao, *Yaowu Fenxi Zazhi*, **14**, 58 (1994).

94. M.H. Vegu, R.F. Saelzer, C.E. Figueroa, G.G. Rios and V.H.M. Jaramillo, *J. Planar Chromatogr. Mod-TLC*, **12**, 72 (1999).

95. A.R. Shalaby, *Food Chem.*, **65**, 117 (1999).

96. H. Furuta, T. Yamane and K. Sugiyama, *J. Chromatogr., Biomed Appl.*, **38**, 103 (1985).

97. K. Haupt, and M.A. Vijayalakshmi, *J. Chromatogr.*, **644**, 289 (1993).

98. A. Ibe, Y. Tamura, H. Kamimura, S. Tabata, H. Hashimoto, M. Iida, and T. Nishima, *Eisei Kagaku.*, **37**, 379 (1991).

99. M. Tod, J.Y. Legendre, J. Chalom, H. Kouwatli, M. Poulou, R. Farinotti and G. Mahuzier, *J. Chromatogr.*, **594**, 386 (1992).

100. K. Saito, F. Yamada, M. Horie and H. Nakazawa, *Anal. Sci.*, **9**, 803 (1993).

101. M.L. Izquierdo Pulido, M.C. Vidal Carou and A. Marine Font, *J. AOAC Int.*, **76**, 1027 (1993).

102. N. Mahy, J. Tussell, and E. Gelpi, *Agents Actions*, **8**, 399 (1978).

103. H. Mita, H. Yasueda and T. Shida, *J. Chromatogr.*, **175**, 339 (1979).

104. F.Y. Lieu and W.G. Jennings, *High Resolut. Chromatogr. Commun.*, **3**, 89 (1980).

105. P.S. Doshi and D.J. Edward, *Life Sci.*, **26**, 1947 (1980).

106. W.F. Staruszkiewicz, jun and J.F. Bond, *J. Assoc. Off. Anal. Chem.*, **64,** 584 (1981).

107. J.J. Keyzer, B.G. Wolthers, H. Breukelman and H.F. Kauffman, *Clin. Chem. Acta*, **121**, 379 (1982).

108. J.J. Keyzer, B.G. Wolthers, H. Breukelman and W. Van der Slik, *J. Chromatogr.*, **275**, 261 (1983).

109. J. Slemr and K. Beyermann, J *Chromatogr.*, **283,** 241 (1984).

110. D.E. Duggan, K.F. Hooke and H.G. Ramijt, *J. Chromatogr. Biomed. Appl.*, **31**, 69 (1984).

111. A. Wollin, H. Navert, *Anal. Biochem.*, **145**, 73 (1985).

112. H. Navert, R. Berube and A. Wollin, *Can. J. Physiol. Pharmacol.*, **63**, 766 (1985).

113. H. Navert, G. Dupuis, and A. Wollin, *J. Chromatogr. Biomed. Appl.*, **56**, 128 (1986).

114. P.L. Rogers and W. Staruszkiewics, *J-AOAC-Int*, **80**, 591 (1997).

115. H. Schmidtlein, *Lebensmittelchem.Gerichtl Chem.*, **33**, 81. (1979).

116. Y. Tsuruta, S. Ishida, K. Kohashi and Y. Ohkura, *Chem. Pharm. Bull.*, **29**, 3398 (1981).

117. E. Gaetani, C.F. Laureri and M. Vitto, *Farmaco, Ed. Prat,* **36**, 496 (1981).

118. G. Skofitsch, A. Saria, P. Holzer and F. Lembeck, *J. Chromatogr. Biomed. Appl.*, **15**, 53 (1981).

119. A. Yoshida, A. Nakamura, *Shokuhin Eiseigaku Zasshi*, **23**, 339 (1982).

120. B. Neidhart and G. Baumhoer, *Fresenius Z Anal. Chem.*, **313**, 564 (1982).

121. K. Volk and H. Gemmer, *Fleischwirtschaft,* **62**, 588 (1982).

122. C.L. Mett and R.J. Sturgeon, *J. Chromalogr.*, **235**, 536 (1982).

123. J.C. Robert, J. Vatier, B.K. Nguyen Phuoc, and S. Bonfils, *J. Chromatogr. Biomed. Appl.*, **24**, 275 (1983).

124. American Society of Brewing Chemists Inc., *J. Am. Soc. Brew. Chem.*, **41**, 110 (1983).

125. U. Pechanek, G. Blaicher, W. Pfannhauser and H. Woidich, *Chromatographia*, **13**, 421 (1980).

126. D. Froehlich and R. Battaglia, *Mtt. Geb. Lebensmittelunters,* **71**, 38 (1980).

127. 1.C.E. Werkhoven, U.A.T. Brinkman and R.W. Frei, *Anal. Chem. Acta,* **114,** 147 (1980).

128. M.L. Romero, L.I. Escobar, X. Lozoya and R.G. Enriquez, *J. Chromatogr.,* **281**, 245 (1983).

129. C. Droz, and H. Tanner, *Schweiz. Z Obst-Weinbau,* **119**, 75 (1983).

130. H.Y. Hui, and S.L. Taylor, *J. Assoc. Off. Anal. Chem.,* **66**, 853 (1983).

131. M.J. Walters, *J. Assoc. Off. Anal. Chem.,* **67**, 1040 (1984).

132. G. Granerus and U. Wass, *Agents, Actions,* **14**, 341 (1984).

133. Y. Nakano, M. Yamaguchi, Y. Tsuruta, Y. Ohkura, T. Aoyama and M. Horioka, *J. Chromatogr. Biomed., Appl.,* **36**, 390 (1984).

134. J.Y. Hui and S.L. Taylor, *J. Chromatogr.,* **312**, 443 (1984).

135. A. Bettero, M.R. Angi, F. Moro and C.A. Benassi, *J. Chromatogr. Biomed. Appl,* **35**, 390 (1984).

136. I. Imamura, K. Maeyama, T. Watanabe and H. Wada, *Anal. Biochem.,* **139**, 444 (1984).

137. A.L. Ronnberg, C. Hansson and R. Hakanson, *Anal. Biochem.,* **139**, 338 (1984).

138. Y. Tonogai, Y. Ito and M. Harada, *Shokuhin Eiseigaku Zasshi,* **25**, 41 (1984).

139. C.F. Laureri, E. Gaetani, M. Vitto and F. Bordi, *Farmaco, Ed. Prat.,* **39**, 29 (1984).

140. C. Buteau, C.L. Duitschaever and G.C. Ashton, *J. Chromatogr.,* **284**, 201 (1984).

141. S.F. Chang, J.W. Ayres and W.E. Sandine, *J. Dairy Sci.,* **68**, 2840 (1985).

142. K.A. Jacobson, T. Marshall, K. Mine, K.L. Kirk and M. Linnoila, *FEBS Lett.,* **188**, 307 (1985).

143. A. Yamatodani, H. Fukuda, H. Wada, T. lwaeda and T. Watanabe, *J. Chromatogr. Biomed. Appl.,* **45**, 115 (1985).

144. J.L. Devalia, B.D. Sheinman and R.J. Davies, *J. Chromatogr. Biomed. Appl.,* **44**, 407 (1985).

145. M.R. Angi, A. Bettero and C.A. Benassi, *Agents Actions,* **16**, 84 (1985).

146. A. Bettero, F. Galiano, C.A. Benassi and M.R. Angi, *Food Chem. Toxicol.,* **23**, 303 (1985).

147. D.L. Ingles, J.F. Back, D. Gallimore, R. Tindale, and K.J. Show, *J. Sci. Food Agric.,* **36**, 402 (1985).

148. S. Allenmark, S. Bergstorm and L. Enerback, *Anal. Biochem.,* **144**, 98 (1985).

149. L.G. Harsing, Jun., H. Nagashima, E.S. Vizi and D. Duncalf, *J. Chromatogr. Biomed. Appl.,* **56**, 19 (1986).

150. Y. Arakawa and S. Tachibana, *Anal. Biochem.,* **158**, 20 (1986).

151. L.G. Harsing, Jun, H. Nagashima, D. Duncalf and E. S. Vizi, *Clin. Chem.,* **32**, 1823 (1986).

152. V.R. Villanueva, M. Mardon and M.T. Le Goff, *Int. J. Environ. Anal. Chem.,* **25**, 115 (1986).

153. K. Mine, K.A. Jacobson, K.L. Kirk, Y. Kitajima, and M. Linnoila, *Anal. Biochem.,* **152**, 127 (1986).

154. R.B.H. Wills, J. Silalahi and M. Wootton, *J. Liq. Chromatogr.,* **10**, 3183 (1987).

155. J.P. Gouygou, C. Sinquin and P. Durand, *J. Food Sci.*, **52**, 925 (1987).

156. A.A. Houdi, P.A. Crooks, G.R. Van-Loon and C.A.W. Schubert, *J. Pharm. Sci.*, **76**, 398 (1987).

157. M.A.J.S. Van Boekel and A.P. Arentsen-Stasse, *J. Chromatogr.*, **389**, 267 (1987).

158. G.R. Pozo, and E.S. Saitua, *Alimentaria* (Mardrid), **25**, 27 (1988).

159. E. Kasziba, L. Flancbaum, J,C. Fitzpatrick and J. Schneiderman, *J. Chromatogr. Biomed. Appl.*, **76**, 315 (1988).

160. S. Karlsson, Z.G. Banhidi and A.C. Albertsson, *J. Chromatogr.*, **442,** 267 (1988).

161. R.E. Schmitt, J. Haas and R. Amado, *Z. Lebensm. Unters. Forsch.*, **187**, 121 (1988).

162. F.V. Carlucci and E. Karmas, *J. Assoc. Off. Anal. Chem.*, **71**, 564 (1988).

163. R. Czerwonka, D. Tsikas and G. Brunner, *Chromatographia*, **25**, 219 (1988).

164. J. Rosier and C. Van Peteghem, *Z. Lebensm. Unters. Forsch*, **186**, 25 (1988).

165. G. Chiavari, G.C. Galletti and P. Vitali, *Chromatographia*, **27**, 216 (1989).

166. A. Lebedzinska, K.I. Eller and V.A. Tutel'yan, *Zh. Anal. Khim.*, **44**, 928 (1989).

167. J.P. Gouygou, C. Martin, C. Sinquin and P. Durand, *Oceanis*, **15**, 599 (1989).

168. S.P. Ashmore, A.H. Thomson and H. Simpson, *J. Chromatogr. Biomed. Appl.*, **88**, 435 (1989).

169. H. Yamanaka, and M. Matsumoto, *Shokuhin Eiseigaku Zasshi*, **30**, 396 (1989).

170.   P.D. Siegel, D.M. Lewis and S.A. Olenchock, *Anal. Biochem.*,
        188, 416 (1990).

171.   U. Buetikofer, D. Fuchs, D. Hurni and J.O. Bosset, *Milt. Geb.*
        Lebensmittelunters Hyg., **81**, 120 (1990).

172.   R. Etter, S. Dietrich, and R. Battazlia, *Mitt. Geb.*
        Lebensmittelunters Hyg., **81**, 106 (1990).

173.   S. Suzuki, K. Kobayashi, J. Noda, T. Suzuki and K. Takama, *J.*
        Chromatogr., **508**, 225 (1990).

174.   Y. Maeno, F. Takabe, H. Inoue, M. Iwasa, *Forensic Sci. Int.*, **46**,
        255 (1990).

175.   L. Leino A. Juhakoski and L. Lauren, *Agents Actions*, **31**, 178
        (1990).

176.   M. Pasto and J. Sabria, *Biomed. Chromatogr.*, **4**, 245 (1990).

177.   G. Cirilli, C.S. Aldana Cirilli and B. Spoettl, *Ind. Aliment.*, **30**, 371
        (1991).

178.   M. Calull, R.M. Marce, J. Fabregas and F. Borrull,
        Chromatographia, **31**, 133 (1991).

179.   M.C. Gennaro and C. Abrigo, *Chromatographia*, **31**, 381 (1991).

180.   H. Fukuda, A. Yamatodani, 1. Imamura, K. Maeyama, and T.
        Watanabe, *J. Chromatogr. Biomed. Appl.*, **105**, 459 (1991).

181.   F.X. Zhou, J. Wahlberg and I.S. Krull, *J. Liq. Chromatogr.*, **14**,
        1325 (1991).

182.   B. Washington, M.O. Smith, T.J. Robinson and R.F. Ochillo, *J.*
        Liq. Chromatogr., **14**, 1417 (1991),

183.   B. Washington, M.O. Smith, T.J. Robinson and J.O. Olubadewo, *J.*
        Liq. Chromatogr., **14**, 2189 (1991).

184.   M. Krizek, *Arch. Anim. Nutr.*, **41**, 97 (1991).

185. C. Tricard, J.M. Cazabeil and M.H. Salagoity, *Analusis,* **19**, M53
      (1991).

186. G.C. Yen, and C.L. Hsieh, *J. Food Sci.,* **56**, 158 (1991).

187. T.A. Neubecker, N.S. Miller, T.N. Asquith and K.E. Driscoll, *J.
      Chromatogr. Biomed. Appl.,* **112**, 340 (1992).

188. S. Moret, R. Bortolomeazzi and G. Lercker, *J. Chromatogr.,* **591**,
      175 (1992).

189. K. Saito, M. Horie, N. Nose, K. Nakagomi, H. Nakazawa, *J.
      Chromatogr.,* **595**, 163 (1992).

190. R. Velasquez, D. Tsikas and G. Brunner, *Fresenius, J. Anal.
      Chem.,* **343**, 78 (1992).

191. P. Lehtonen, M. Saarinen, M. Vesanto and M.L. Riekkola, *Z.
      Lebensm. Unters. Forsch.,* **194**, 434 (1992).

192. R. Cozzani, A.L. Cinquina, and D. Barchi, *Ind. Aliment,* **31**, 1000
      (1992).

193. G.L. Huang, P. Li and P.Y. Pu, *Sepu,* **11**, 37 (1993).

194. B. Washington, K. Nguyen and R.F. Ochillo, *J. Liq. Chromatogr.,*
      16, 1195 (1993).

195. D. Tsikas, R. Velasquez, C. Tolenado and G. Brunner, *J.
      Chromatogr. Biomed. Appl.,* **125**, 37 (1993).

196. C.M.C.J. Van Haaster, W. Engels, P.J.M.R. Lemmens, G.
      Hornstra, and G.J. Van der Vusse, *J. Chromatogr. Biomed. Appl.,*
      128, 233 (1993).

197. K. Maeyama, M. Sasaki, and T. Watanabe, *Anal. Biochem.,* **194**,
      316 (1991).

198. D.N. Xie, Y. Chen and Z.G. Guo, *Sepu.,* **12**, 65 (1994).

199. M. Katayama, H. Takeuchi, and H. Taniguchi, *Anal. Chem. Acta.,*
      287, 83 (1994).

200.   G. Achilli, G.P. Cellerino and G. Melzi-d'Eril, *J. Chromatogr.*, **661**, 201 (1994).

201.   K. Kuruma, E. Hirai, K. Uchida, J. Kikuchi and Y. Terui, *Anal. Sci.*, **10**, 259 (1994).

202.   O. Busto, Y. Valero, J. Guasch and F. Borrull, *Chromatographia*, **38**, 571 (1994).

203.   M. Nakazato, K. Saito, S. Morozumi, T. Wauke, F. Ishikawa, K. Fujinuma, T. Moriyasu, T. Nishima and Y. Tamura, *Jpn J. Toxicol. Environ. Health*, **40**, 203 (1994).

204.   G. Kalligas, I. Kaniou, G. Zachariadis, H. Tsoukali and P. Epivatianos, *J. Liq. Chromatogr.*, **17**, 2457 (1994).

205.   D. Lowe-Walters, J.E. James, F.B. Vest and H.T. Karnes, *Biomed. Chromatogr.*, **8**, 207 (1994).

206.   C.M.C.J. van-Haaster, W. Engles, P.J.M.R. Lemmens, G. Hornstra and G.J. Van-des-Vusse, *J. Chromatogr. Biomed. Appl.*, **657**, 261 (1994).

207.   D.R. Lowe, C. March, J.E. James and H.T. Karnes, *J. Liq. Chromatogr.*, **17**, 3563 (1994).

208.   D. Hornero-Mendez and A. Garrido-Fernandez, *Analyst*, **119**, 2037 (1994).

209.   D. Egger, G. Reisbach and L. Hueltner, *J. Chromatogr., Biomed. Appl.*, **662**, 103 (1994).

210.   K. Hermann, G. Frank and J. Ring, *J. Liq. Chromatogr.*, **18**, 189 (1995).

211.   F. Tarrach, *Dtsch. Lebensm. Rundsch.*, **91**, 73 (1995).

212.   M.K. Alam, M. Sasaki, T. Watanabe and K. Maeyama, *Anal. Biochem.* **229**, 26 (1995).

213.   T.B. Jensen and P.D. Marley, *J. Chromatogr. Biomed. Appl.*, **670**, 199 (1995).

214. M. Sato, T. Nakano, M. Takeuchi, T. Kumagai, N. Kanno, E. Nagahisa and Y. Sato, *Biosci. Biotechnol. Biochem.*, **59**, 1208 (1995).

215. K. Pihel, S.C. Hsieh, J.W. Jorgenson and R.M. Wightman, *Anal. Chem.*, **67**, 4514 (1995).

216. K. Saito, M. Horie and H. Nakazawa, *Shokuhin Eiseigako. Zasshi.*, **36**, 639 (1995).

217. O. Busto, J. Guasch and F. Borrull, *J. Chromatogr.*, **718**, 309 (1995).

218. N. Bilic, *J. Chromatogr.*, **719**, 321 (1996).

219. Y. Yokoyama, O. Ozaki and H. Sato, *J. Chromatogr.*, **739**, 333 (1996).

220. T. Hernandez-Joves, M. Izquierdo-Pulido, M.T. Veciana-Nagues, and M.C. Vidal-Carou. *J. Agric. Food Chem.*, **44**, 2710 (1996).

221. S. Moret, L.S. Conte and F. Callegarin, *Ind. Aliment.*, **35**, 650 (1996).

222. D.F. Hwang, S.H. Chang, C.Y. Shiua and T.J. Chai, *J. Chromatogr. Biomed. Appl.*, **693**, 23 (1997).

223. J. Kirschbaum, I. Busch and H. Brueckner, *Chromatographia*, **45**, 263 (1997).

224. K.W. Lee and Y.C. Lee, *Anal. Sci. Technol*, **10**, 43 (1997).

225. K. Saito, M. Horie, Y. Tokumaru, S. Hattori and H. Nakazawa, *Shokuhin. Eiseigaku. Zasshi*, **39**, 39 (1998).

226. O. Vandenabeele, L. Garelly, M. Ghelfenstein, A. Commeyras and L. Mion, *J. Chromatogr.*, **795**, 239 (1998).

227. R. Pacheco-Aguilar, M.E. Lugo-Sanchez, R.E. Villegas-Ozuna and R. Robles-Burgueno, *J. Food Compos. Anal.*, **11**, 188 (1998).

228. V. Frattini and C. Lionette, *J. Chromatogr.*, **809**, 241 (1998).

229. E. Kinoshita and M. Saito, *Biosc. Biotechnol. Biochem.*, **62**, 1488 (1998).

230. M. Arlorio, J.D. Coisson and A. Martelli, *Chromatographia*, **48**, 763 (1998).

231. W.D. Liu, W. Qi and H.L. Zheng. *Seup.* **17**, 80 (1999).

232. O. Aygun, E. Schneider, R. Scheuer, E. Usleber, M. Gareis and E. Martlbauer, *J. Agric. Food Chem.*, **47**, 1961 (1999).

233. H. Fujino and S. Goya, *Yakugaku Zasshi*, **110**, 693 (1990).

234. U. Pechanek, G. Blaicher, W. Pfannhauser and H. Woidich, *Z. Lebensm. Unters Forsch.*, **171**, 420 (1980).

235. S.J. Lewis, M.R. Fennessy, F.J. Laska and D.A. Taylor, *Agents Actions*, **10**, 197 (1980).

236. R.P. Andrews and N.A. Baldar, *Sci. Tools*, **30**, 8 (1983).

237. R. Draisci, S. Cavalli, L. Lucentini and A. Stacchini, *Chromatographia*, **35**, 584 (1993).

238. J.A. Zee, R.E. Simard and L.L. Heureux, *Lebensm. Wiss. Technol.*, **18**, 245 (1985).

239. R. Pfeiffer, H.G. Greulich and H. Erbersdobler, *Lebensmittelchem. Gerichtl. Chem.*, **40**, 118 (1986).

240. L. Simon-Sarkadi and W.H. Holzapfel, *Z. Lebensm-Unters-Forsch.*, **198**, 230 (1994).

241. E. Trevino, D. Beil and H. Steinhart, *Food Chem.*, **58**, 385 (1997).

242. L.B. Hough, P.L. Stetson and E.F. Domino, *Anal. Biochem.*, **96**, 56 (1979).

243. H. Mita, H. Yasueda and T. Shida, *J. Chromatogr. Biomed. Appl.*, 153 (1980).

244. H. Mita, H. Yasueda and T. Shida, *J. Chromatogr. Biomed. Appl.*, **10**, 1 (1980).

245. H. Mita, H. Yasueda and T. Shida, *Koenshu Iyo Maso Kenkyukai,* **5**, 169 (1980).

246. L.B. Hough, H.K. Khandelwal, A.M. Morrishow and J.P. Green, *J. Pharmacol. Methods,* **5**, 143 (1981).

247. J.D. Henion, J.S. Nosanchuk and B.M. Bilder, *J. Chromatogr.,* **213**, 475 (1981).

248. J.J. Keyzer, B.G. Wolthers, F.A.J. Muskiet, H.F. Kauffman and A. Groen, *Clin. Chem. Acta,* **113**, 165 (1981).

249. C.G. Swahn and G. Sedvall, *J. Neurochem,* **37**, 461 (1981).

250. C.G. Swahn and G. Sedvall, *J. Neurochem.,* **40**, 688 (1983).

251. L.J. Roberts II and J.A. Oates, *Anal. Biochem.,* **136**, 258 (1984).

252. L.J. Roberts, II, *J. Chromatogr.,* **287**, 155 (1984).

253. J.J. Keyzer, B.G. Wolthers, F.A.J. Muskiet, H. Breukelman, H.F. Kauffman and K. De Vries, *Anal. Biochem.,* **139**, 474 (1984).

254. L.J. Roberts, II; K.A. Aulsebrook and J.A. Oates, *J. Chromatogr., Biomed. Appl.,* **39**, 41 (1985).

255. J.J. Keyzer, H. Breukelman, B.G. Wolthers, F.J. Richardson and J.G.R. De Monchy, *Agents Actions,* **16**, 76 (1985).

256. N.A. Payne, A. Zirrolli and G. Gerber, *Anal. Biochem.,* **178**, 414 (1988).

257. S. Murray, G. O'Malley, I.K. Taylor, A.I. Mallet and G.W. Taylor, *J. Chromatogr. Biomed. Appl.,* **83**, 15 (1989).

258. S. Murray, R. Wellings, I.K. Taylor, R.W. Fuller and G.W. Taylor, *J. Chromatogr. Biomed. Appl.,* **105**, 289 (1991).

259. A.B. Einarsson, and U. Moberg, *J. Chromatogr.,* **209**, 121 (1981).

260. K. Rubach, P. Offizorz and C. Bryer, *Z Lebensm. Unters. Forsch.,* **172**, 351 (1981).

261.   R. Karlsson and R.R. Einarsson, *Anal. Lett.,* **15**, 909 (1982).

262.   R. Jarofke, *J. Chromatogr.,* **390**, 161 (1987).

263.   K. Priebe, *Fleischwirtschaft,* **59**, 1658 (1979).

264.   H. Kikuchi and M. Watanabe, *Anal. Biochem.,* **115**, 109 (1981).

265.   M.Y. Kamel and S.A. Maksoud, *J.* Chromatogr., **283**, 331 (1984).

266.   Chiari, M. Giacomini, C. Micheletti and P.G. Righetti, *J. Chromatogr.,* **558**, 285 (1991).

267.   A. Nann, I. Silvestri and W. Simon, *Anal. Chem.,* **65**, 1662 (1993).

268.   K.D. Altria, P. Frake, I. Gill, T. Hadgett, M.A. Kelly and D.R. Rudd, *J. Pharm. Biomed. Anal.*, **13**, 951 (1995).

269.   C. Bruce, W.H. Taylor and A. Westwood, *Ann. Clin. Biochem.,* **16**, 259 (1979).

270.   C. Dent, F. Nilam and I.R. Smith, *Agents Actions,* **9**, 34 (1979).

271.   L. Guilloux, D. Hartmann and G. Ville, *Clin. Chem. Acta.,* **116**, 269 (1981).

272.   K.M. Verburg, R.R. Bowsher and D.P. Henry, *Life Sci.,* **32**, 2855 (1983).

273.   K. Warren, J. Dyer, S. Merlin and M. Kaliner, *J. Allergy Clin. Immunol.,* **71**, 206 (1983).

274.   R.J. Harvima, H. Neittaanmaki, I.T. Harvima, E.D. Kajander, and J.E. Fraki, *Clin. Chim. Acta,* **143**, 337 (1984).

275.   J.K. Brown, M.J. Frey, B.R. Reed, A.R. Leff, R. Shields and W.M. Gold, *J. Allerg. Clin Immunol.,* **73**, 473 (1984).

276.   M. Haimart, J.M. Launay, G. Zuercher, N. Cauet, C. Dreux and M. Da Prada, *Agents Actions,* **16**, 71 (1985).

277.   D. Chevrier, J.L. Guesdon, J.C. Mazie and S. Avrameas, *J. Immunol. Methods,* **94**, 119 (1986).

278. D.O. Rauls, S. Ting and M. Lund, *J. Allerg. Clin. Immunol.*, **77**, 673 (1986).

279. J.L. Guesdon, D. Chevrier, J.C. Mazie, B. David and S. Avrameas, *J. Immunol. Methods*, **87**, 69 (1986).

280. R.J. Harvima, I.T. Harvima and J.E. Fraki, *Clin. Chem. Acta*, **171**, 247 (1988).

281. A Morel, M. Darmon and M. Delaage, *Agents Actions*, **30**, 291 (1990).

282. J. Cepicka, P. Rychetsky, I. Hochel, F. Strejcek and P. Rauch, *Monatsscher. Brauwiss.*, **48**, 161 (1992).

283. P. Rauch, P. Rychetsky, 1. Hochel, R. Bilek and J.L. Guesdon, *Food Agric. Immunol.*, **4**, 67 (1992).

284. L.R. Hegstrand, *Biochem. Pharmacol.*, **34**, 3711 (1985).

285. K. Painter and C.R. Vader, *Clin. Chem.*, **25**, 797 (1979).

286. L.M. Peyret, P. Moreau, J. Dulluc and M. Geffard, *J. Immunol. Methods*, **90**, 39 (1986).

287. F. Nambu, S. Murakata, T. Shiraji, N. Omawari, M. Sawada, T. Okegawa, A. Kaswasaki and S. Ikeda, *Prostaglandins*, **39**, 623 (1990).

288. E. Oosting and J.J. Keyzer, *Agents Actions*, **33**, 215 (1991).

289. R.A. Erger and T.B. Casale, *J. Immunol. Methods*, **152**, 115 (1992).

290. E. Hammar, A. Berglund, A. Hedin, A. Norrman, K. Rustas, U. Ytterstrom, and E. Akerblom, *J. Immunol. Methods*, **128**, 51 (1990).

291. M.J. Brown, P.W. Ind, P.J. Barnes, D.A. Jenner and C.T. Dollery, *Anal. Biochem.*, **109**, 142 (1980).

292. D. Serrar, R. Berbant, S. Bruneau, and G.A. Denoyel, *Food Chem.*, **54**, 85 (1995).

293. C. Krueger, U. Swing, G. Stengel, I. Kema, J. Westermann and B. Mans, *Arch-Lebensmittlhyg.*, **46**, 115, 118 (1995).

294. C. Weise, D. Huebner, and K. Speer, *Lebensmittlchemie*, **51**, 118 (1997).

295. J. Dyer, K. Warren, S. Merlin, D.D. Metcalfe and M. Kaliner, *J. Allergy Clin. Immunol.*, **70**, 82 (1982).

296. M. Friedman and A.T. Noma, *J. Chromatogr.*, **219**, 343 (1981).

297. G.J. Gleich and W.M. Hull, *J. Allergy Clin. Immunol*, **66**, 295 (1980).

298. J. Murray and A.S. McGill, *J. Assoc. Off. Anal. Chem.*, **65**, 71 (1982).

299. R.K. Marwaha, B.F. Johnson and G.E. Wright, *Am. J. Hosp. Pharm.*, **42**, 1568 (1985).

300. M.R. Pratter, R.K. Marwaha, R.S. Irwin, B.F. Johnson and F.J. Curley, *Am. Rev. Respir. Dis.* **132**, 1130 (1985).

301. R.K. Marwaha and B.F. Johnson, *Am. J. Hosp. Pharm.*, **43**, 380 (1986).

302. N.H. Nielsen, F. Madsen, L. Frolund, V.G. Svendsen and B. Weeke, *Allergy*, **43**, 454 (1988).

303. C. McDonald, J.F. Parkin, C.A. Richardson, M. Sweidan, D. Lonergan, C. Chan and R. Cohen, *J. Clin. Pharm. Ther.*, **15**, 41 (1990).

304. P. Marshik, S. Moghaddam, I. Tebbett and L. Hendeles, *Chest*, **115**, 194 (1999).

IBUPROFEN

John D. Higgins,[1] Timothy P. Gilmor,[1]
Stephan A. Martellucci,[1] Richard D. Bruce[1]

and

Harry G. Brittain[2]

(1) McNeil Consumer Healthcare
7050 Camp Hill Road
Fort Washington, PA 19034
USA

(2) Center for Pharmaceutical Physics
10 Charles Road
Milford, NJ 08848
USA

265

Contents

1.   Description

1.1   Nomenclature

1.1.1   Chemical Name

(±)-2-(4-*iso*butylphenyl) propionic acid

(±)-(α-methyl-4(2-methyl-propyl)benzeneacetic acid

p-isobutyl-hydratropic acid

1.1.2   Nonproprietary Names

rac-ibuprofen

1.1.3   Proprietary Names

Advil; Brufen; Ibufen; Motrin; Nuprin; Nurofen; Paduden; Proflex; Rufin; Unipron

1.2   Formulae

1.2.1   Empirical

$C_{13}H_{18}O_2$

1.2.2   Structural

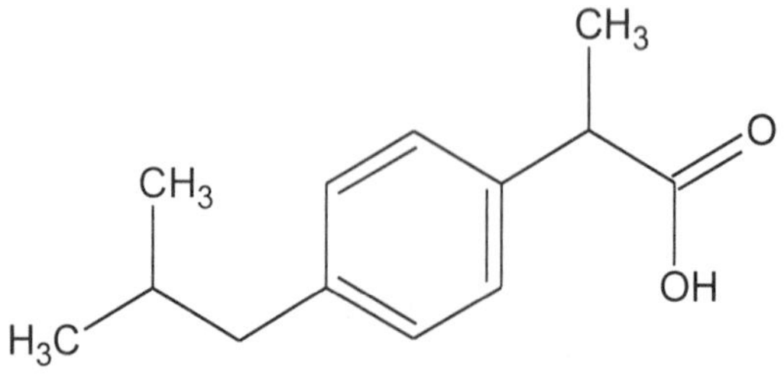

1.3   Molecular Weight and CAS Number

MW = 206.281
CAS =15687-27-1

1.4   Appearance

White powder or crystals

1.5   Uses and Applications

Rac-ibuprofen was introduced in the late sixties as a safe non-steroidal anti-inflammatory drug (NSAID) for the treatment of a wide range of indications, including pain, inflammation, arthritis, fever and dysmenorrhea [1]. While *rac*-ibuprofen has been shown to be relatively safe in doses ranging up to 2400 mg/day [2], typical effective oral doses range from 600-1800 mg/day [1].

It is well established that the *S*-(+) enantiomer is almost entirely responsible for the anti-inflammatory effects of *rac*-ibuprofen [3]. The main mechanism of action is known to be the inhibition of prostanoid biosynthesis via blockade of cyclo-oxygenase (COX) [4]. The COX enzyme has been shown to exist as two isoforms. COX-1 is a constitutive protein present in a wide range of cells, and is important in the regulation of prostaglandins that are involved in the protection of the lining of the GI tract from noxious agents. It is thought that inhibition of COX 1 contributes to gastric ulceration, which is one of the most frequent side effects of NSAID therapy. COX-2 is the inducible form of the enzyme that is expressed in macrophages and other immunoregulatory cells after trauma.

Hence, the COX-2 isoform plays an important role in inflammatory processes. Although ibuprofen has been found to inhibit both COX isoforms *in vitro*, it still displays an excellent safety profile with a relatively low incidence of GI toxicity observed at clinical doses. These findings, along with a discussion of the physiologic activities of the enantiomerically pure (*R*)- and (*S*)-forms of ibuprofen, have been recently reviewed [3].

2.   <u>Methods of Preparation</u>

A myriad of routes for the synthesis of *rac*-ibuprofen, and the separated enantiomers, have been reported [5]. One of the earliest industrially feasible routes was developed by the Boots Pure Drug Company [6], and is shown in Scheme 1. The synthesis starts with the Al(III) catalyzed acylation of isobutyl benzene (**2**) to form 4-*iso*butylacetophenone (**3**). Darzen reaction of (**3**) with ethyl chloroacetate and sodium ethoxide affords the epoxyester (**4**). Hydrolysis and decarboxylation of (**4**) gives the aldehyde (**5**), which is then oxidized to produce *rac*-ibuprofen in good yield and purity.

J.D. HIGGINS, T.P. GILMOR, S.A. MARTELLUCCI,
R.D. BRUCE, AND H.G. BRITTAIN

Scheme 1. Early industrial synthesis of ibuprofen.

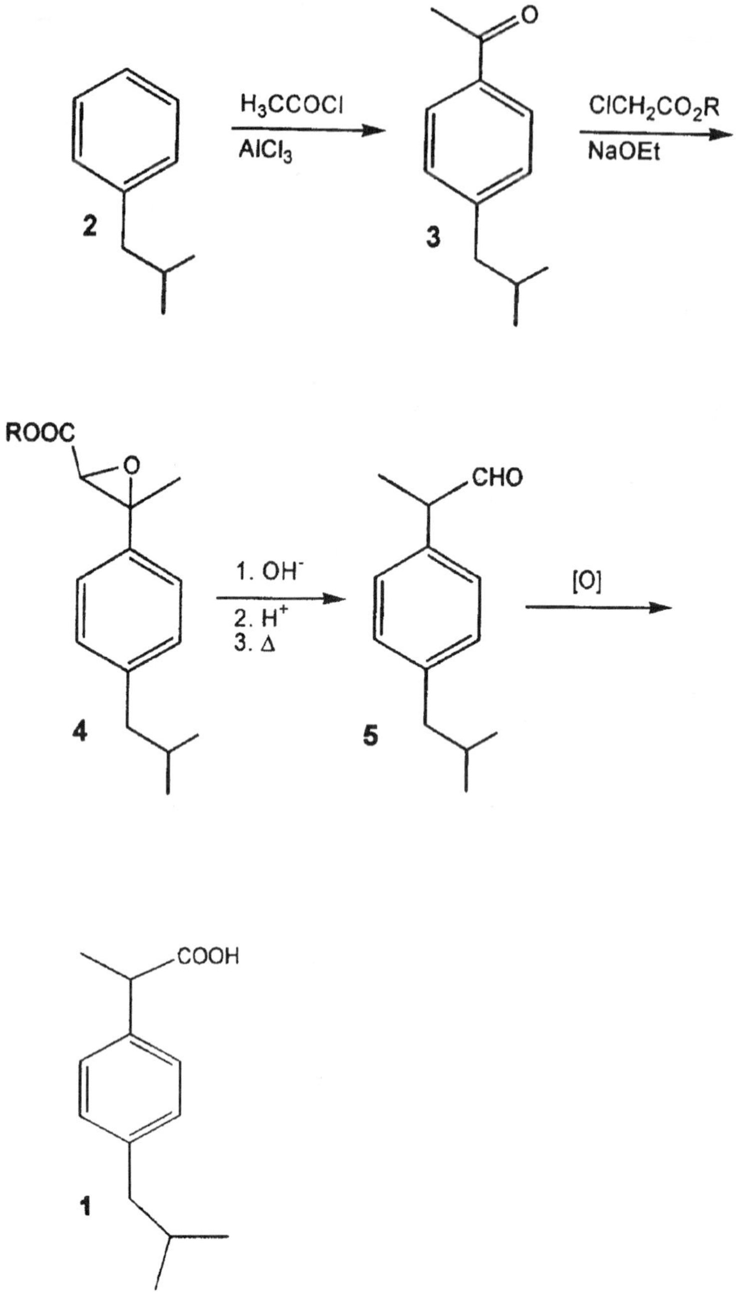

More recently, a shorter three step catalytic route has been developed [7], and is illustrated in Scheme 2. Here, a Pd catalyzed carbonylation reaction is employed in the final step to introduce the carboxylate group.

Scheme 2.   Carbonylation route to ibuprofen.

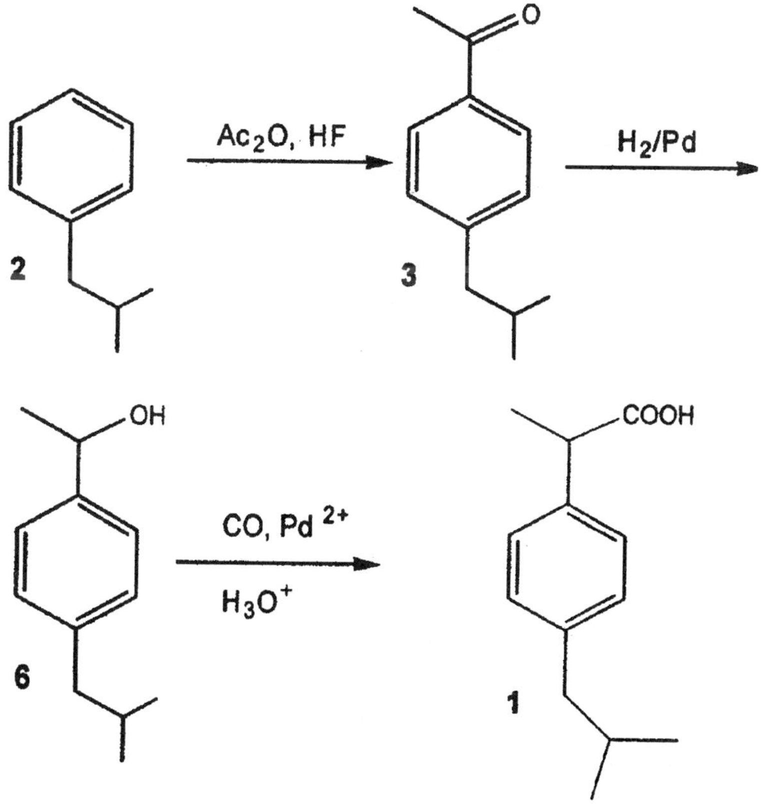

3.   **Physical Properties**

3.1   **Crystallographic Properties**

3.1.1   **Single Crystal Structure**

In an early report, the rac-ibuprofen was found to crystallize in the
monoclinic space group $P2_1/c$ [8], and the unit cell parameters were
reported as:

a:	14.667 Å
b:	7.886 Å
c:	10.730 Å
β:	99.362°
Z:	4 molecules per unit cell

In a subsequent study, crystals of rac-ibuprofen suitable for single crystal
X-ray diffraction analysis were isolated from an acetonitrile solution [9].
The same monoclinic space group ($P2_1/c$) was deduced, and the other
relevant crystal data from this study are as follows:

a:	14.397(8) Å
b:	7.818(4) Å
c:	10.506(6) Å
β:	99.70(3)°
Z:	4 molecules per unit cell

A complete study of the conformational flexibility of ibuprofen was
undertaken by searching crystallographic databases, and conducting
potential energy calculations [10]. The preferred conformation of the
substance was deduced from this work, and is shown in Figure 1.

3.1.2   **Polymorphism**

Other than its tendency to adopt a variety of particle morphologies [11-13],
rac-ibuprofen does not appear to exhibit genuine polymorphism.
Additionally, although polymorphs of rac-ibuprofen have not been
reported, the lysine salt of rac-ibuprofen has been shown to exist in two
polymorphic forms [14].

Figure 1.   Optimized conformation of *rac*-ibuprofen [10].

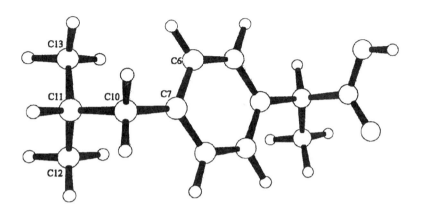

It should be noted that *rac*-ibuprofen materials manufactured by different sources can have different crystal dislocations, surface rugosity and/or surface areas. Although these differences do not significantly effect the melting point, they can influence compression and granulation processes [15]. Studies on the effects of ibuprofen crystallinity on tablet formulation and processing also have been described. For example, it has also been reported that although polymorphic changes were not observed upon compaction, subtle crystal lattice modifications in ibuprofen can occur which ultimately may have effects on dissolution behavior [11].

3.1.3   X-Ray Powder Diffraction Pattern

The x-ray powder diffraction pattern for *rac*-ibuprofen is available either in the ICDD database [16] or in the literature [17]. The database pattern is illustrated in Figure 2.

3.2   Optical Activity

Being a racemic mixture, *rac*-ibuprofen cannot exhibit optical activity. However, the racemate can be easily separated into its component enantiomers, and a number of reports have issued regarding methods of resolution [18-20].

J.D. HIGGINS, T.P. GILMOR, S.A. MARTELLUCCI,
R.D. BRUCE, AND H.G. BRITTAIN

Figure 2.   X-ray powder diffraction pattern of *rac*-ibuprofen [16].

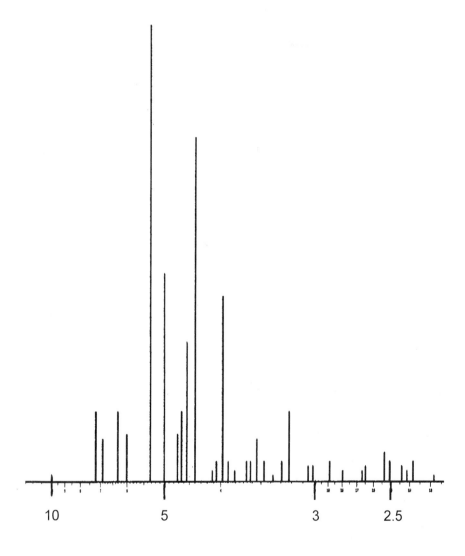

d-Spacing (Å)

The full phase diagram of *rac*-ibuprofen has been determined, where it was established that the substance is crystallized as a racemate solid and not as a conglomerate mixture [21].

3.3. Thermal Methods of analysis

3.3.1 Melting Behavior

Rac-ibuprofen exists as a stable crystalline solid, and exhibits a typical melting range of 75-77°C [22]. Interestingly, if the molten solid is allowed to cool from the melting point to room temperature without vibration in a smooth-lined container, *rac*-ibuprofen can exist as an oil phase for several hours to a few days. If disturbed, an exothermic recrystallization proceeds and a bulk crystal rapidly grows vertically out of the oil phase [23].

The combination of low melting point and slow recrystallization kinetics should be considered when formulating the substance, since even minor frictional heating during compression can melt the material. Furthermore, *rac*-ibuprofen has been shown to have a propensity toward sublimation, which also should be considered when designing manufacturing processes or stability protocols [24].

Finally, it should also be noted that *rac*-ibuprofen is known to form eutectic mixtures with a variety of excipients, which can result in a significant lowering of the melting point [25].

3.3.2 Differential Scanning Calorimetry

A DSC thermogram of *rac*-ibuprofen is shown in Figure 3 [23]. The sample studied exhibited a single melting endotherm, characterized by an onset temperature of 74.0°C and a peak maximum of 75.6°C. The sample was noted to melt without decomposition, and the enthalpy of fusion associated with the melting transition was determined to be 77.2 J/g.

3.3.3 Thermogravimetry

The TG thermogram of *rac*-ibuprofen is also shown in Figure 3 [23]. Owing to the tendency of the compound to undergo sublimation, the weight loss that initiates above 100°C cannot be assigned exclusively to thermal decomposition.

J.D. HIGGINS, T.P. GILMOR, S.A. MARTELLUCCI,
R.D. BRUCE, AND H.G. BRITTAIN

Figure 3. Differential scanning calorimetry and thermogravimetry thermograms of *rac*-ibuprofen [23].

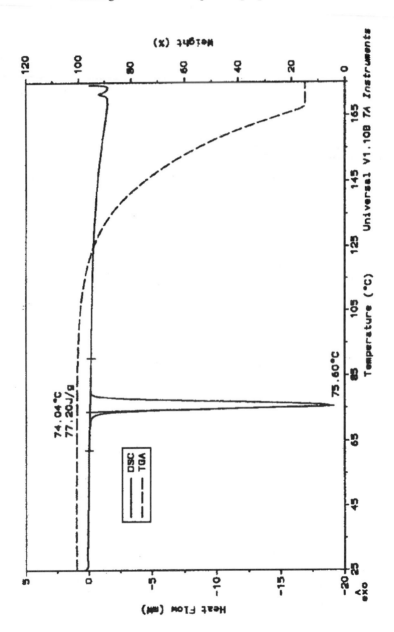

3.4   Ionization Constants

The pKa of ibuprofen has been determined by potentiometric titration in mixed organic/aqueous solvent systems to be in the range of 4.5 to 4.6 [26].

Using the ACD/PhysChem computational program, the pKa of ibuprofen was calculated to be 4.41 ± 0.20 [27], in good agreement with the empirically determined value.

3.5   Solubility Characteristics

The solubility of *rac*-ibuprofen in a number of non-aqueous solvents is compiled in Table 1 [28]. The substance is readily soluble in alcohols, chlorinated hydrocarbon solvents, and dimethyl sulfoxide, but is only sparingly soluble in nonpolar hydrocarbon solvents.

In the aqueous conditions also summarized in Table 2, *rac*-ibuprofen is seen to be nearly insoluble at low pH, but readily soluble in basic media [23].

The solubility of ibuprofen is also affected by stereochemistry. At pH 1.5 *rac*-ibuprofen has an aqueous solubility of 4.6 mg/100 mL, whereas the enantiomeric R(-) and S(+) forms have solubilities of 9.6 and 9.5 mg/100 mL, respectively [29].

A more detailed evaluation of the solubility of *rac*-ibuprofen was obtained using the ACD/PhysChem computational program [27]. The results of this calculation are located in Figure 4. The agreement between the calculated values and the empirical data is seen to be quite good.

3.6   Partition Coefficients

The pH dependence of octanol/water distribution coefficients of *rac*-ibuprofen was obtained using the ACD/PhysChem computational program [27]. The results of this calculation are found in Figure 5.

Table 1

Solubility of *rac*-Ibuprofen in Various Solvent Systems

Solubility of *rac*-Ibuprofen in Non-Aqueous Solvents (at 20°C)

Solvent System	Approximate Solubility (% w/v)
Liquid paraffin	< 0.1
Heptane	2.3 – 2.6
Hexane	3.3 – 3.6
Petroleum ether	2.3 – 2.6
Chloroform	65 – 70
Ethyl alcohol (dehydrated)	60 – 70
Isopropyl alcohol	30 – 32
N-Octanol	20 – 22
Propylene glycol	20 – 25
Ethylene glycol	2 – 2.5
Acetone	60 – 65
Industrial methylated spirit	50 – 55
Methyl alcohol	60 – 70
N,N-Dimethyl acetamide	60 – 70
Dimethyl sulfoxide	55 – 60
Polyethylene glycol 300	<0.1

Solubility of *rac*-Ibuprofen in Aqueous Systems (at 20°C)

System Composition	Solubility (mg/mL)
DI water (pH 3.0)	<0.1
PH 1 (HCl)	<0.1
PH 4 (phosphate buffer)	<0.1
PH 6 (phosphate buffer)	1.0
PH 8 (phosphate buffer)	>100

Figure 4.   Solubility of *rac*-ibuprofen, as predicted by ACD/PhysChem [27]. The open circles represent the solubility multiplied by 100.

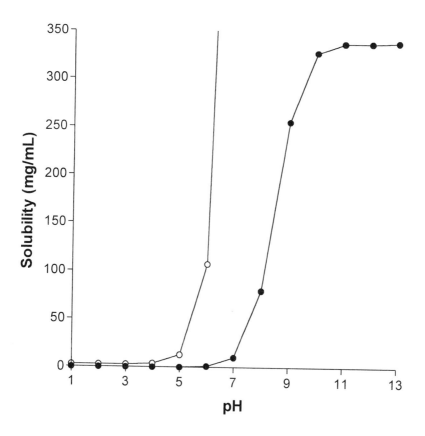

pH	Solubility (mg/mL)	pH	Solubility (mg/mL)
1.0	0.0270	8.0	79.6000
2.0	0.0270	9.0	255.0000
3.0	0.0280	10.0	327.0000
4.0	0.0370	11.0	337.0000
5.0	0.1300	12.0	337.0000
6.0	1.0700	13.0	338.0000
7.0	10.1000		

Figure 5.   Octanol-water distribution coefficients of *rac*-ibuprofen, as
             predicted by ACD/PhysChem [27].

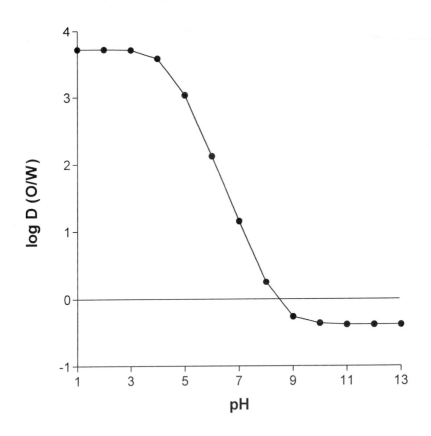

pH	log D	pH	log D
1.0	3.7200	8.0	0.2500
2.0	3.7200	9.0	-0.2600
3.0	3.7100	10.0	-0.3600
4.0	3.5800	11.0	-0.3800
5.0	3.0300	12.0	-0.3800
6.0	2.1200	13.0	-0.3800
7.0	1.1500		

3.7   Spectroscopy

3.7.1   UV/VIS Spectroscopy

The ultraviolet absorption spectrum of *rac*-ibuprofen was obtained in
methanol and in 0.1 N NaOH [23], and these are shown in Figure 6. The
spectrum consists of the weak absorption bands of the phenyl ring that are
found in the 255-275 nm range, for which the molar absorptivity is
approximately 250 L / mol · cm. The other strong feature consists of the
much more intense band system centered around 225 nm, for which the
molar absorptivity is approximately 9000 L / mol · cm.

3.7.2   Vibrational Spectroscopy

The infrared absorption spectrum of *rac*-ibuprofen was obtained in a KBr
pellet [23], and is shown in Figure 7.

3.7.3   Nuclear Magnetic Resonance Spectrometry

3.7.3.1 ^{1}H-NMR Spectrum

The ^{1}H-NMR spectrum of *rac*-ibuprofen was obtained at a frequency of 400
MHz in deuterated dimethyl sulfoxide [23]. The resulting spectrum is
shown in Figure 8 along with assignments for the resonance bands.

3.7.3.2 ^{13}C-NMR Spectrum

The ^{13}C-NMR spectrum of *rac*-ibuprofen was obtained at a frequency of
100 MHz in deuterated dimethyl sulfoxide [23], and the resulting spectrum
is shown in Figure 9.

3.7.4   Mass Spectrometry

The electron-impact mass spectrum [30] of *rac*-ibuprofen is shown in
Figure 10.

Figure 6.   Ultraviolet absorption spectrum of *rac*-ibuprofen in
             methanol and in 0.1 N NaOH [23].

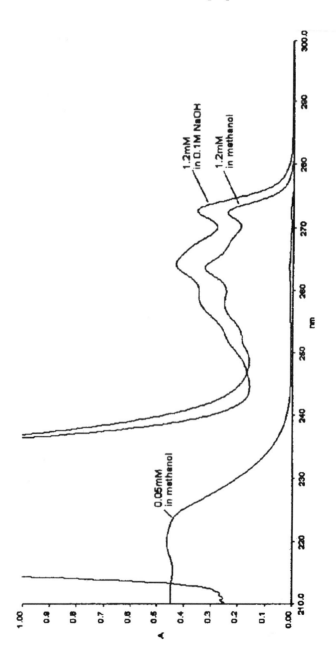

Figure 7.   Infrared absorption spectrum of *rac*-ibuprofen in a KBr
            pellet [23].

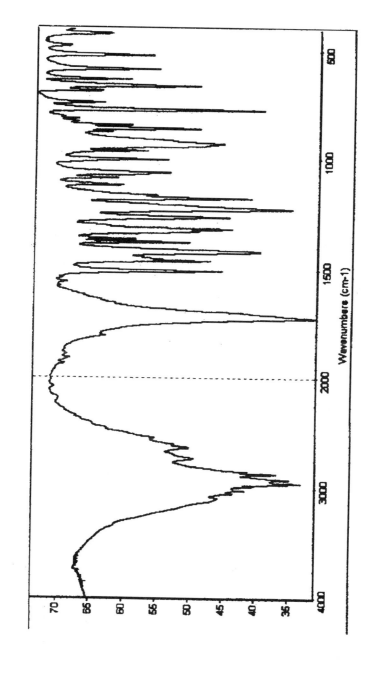

Figure 8. ^{1}H-NMR spectrum of *rac*-ibuprofen, obtained at a frequency
             of 400 MHz in deuterated dimethyl sulfoxide [23].

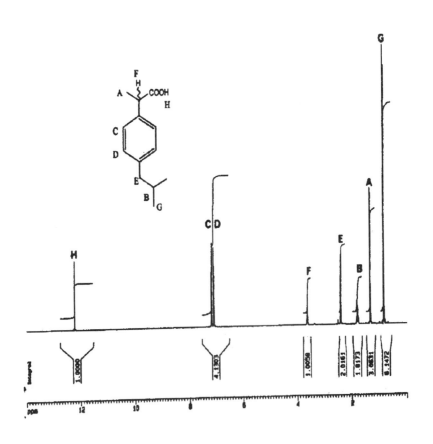

Figure 9. ^{13}C-NMR spectrum of *rac*-ibuprofen, obtained at a
            frequency of 400 MHz in deuterated dimethyl sulfoxide [23].

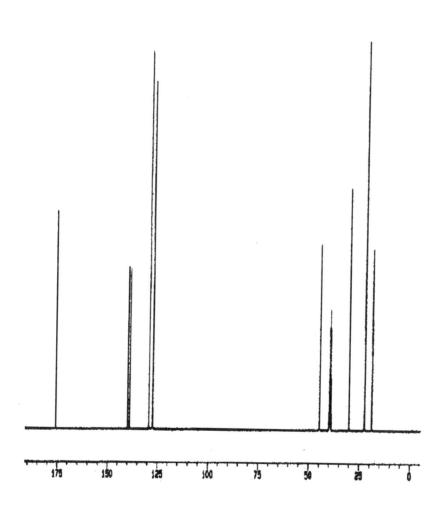

Figure 10.   Electron-impact mass spectrum of *rac*-ibuprofen [30].

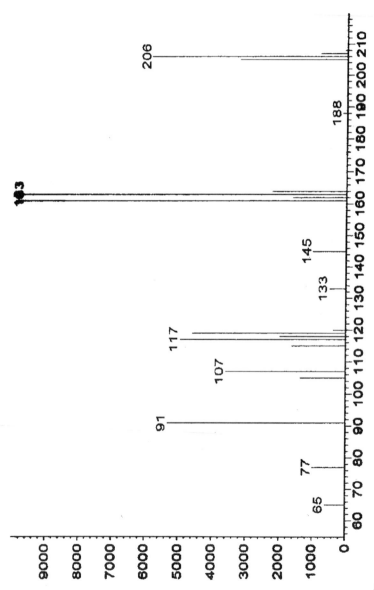

4.   Methods of Analysis

4.1   United States Pharmacopoeia Compendial Tests

USP 24 specifies a number of compendial tests for the drug substance [31]. The test methods, and their specifications, are summarized as follows:

4.1.1   Identification

Test A: Infrared Absorption, according to general test <197M>.

Test B: Ultraviolet Absorption, according to general test <197U>. The test solution is prepared at a concentration of 250 µg per mL, in 0.1 N NaOH. The respective absorptivities at 264 nm and 273 nm, calculated on the anhydrous basis, do not differ by more than 3.0% with respect to the standard.

Test C: The chromatogram of the assay preparation obtained as directed in the Assay Method exhibits a major peak for ibuprofen, the retention time of which (relative to that of the internal standard) corresponds to that exhibited in the chromatogram of the Standard preparation obtained as directed in the Assay.

4.1.2   Water

Perform according to general method <921>, Method I. The test article does not contain more than 1.0%.

4.1.3   Residue on Ignition

Perform according to general method <281>. The test article does not yield more than 0.5%.

4.1.4   Heavy Metals

Perform according to general method <921>, Method II. The test article does not contain more than 0.002%.

4.1.5   Organic Volatile Impurities

Perform according to general method <467>, Method V, using dimethyl sulfoxide as the solvent. The test article meets the requirements.

4.1.6 Chromatographic Purity

The chromatographic purity of the test article is established using high-performance liquid chromatography. The mobile phase is prepared using a suitably filtered mixture of water (previously adjusted with phosphoric acid to a pH of 2.5) and acetonitrile in the ratio of 1340:680. The test preparation consists of a solution of ibuprofen in acetonitrile, containing about 5 mg per mL. The resolution solution is a solution in acetonitrile containing about 5 mg/mL of Ibuprofen and 5 mg/mL of valerophenone. The liquid chromatograph is equipped with a 214-nm detector and a 4-mm x 15-cm column that contains 5-um packing L1, and is maintained at 30 ± 0.2 °C. The flow rate is about 2 mL per minute.

Chromatograph a series of 5-μL injections of the Test Preparation to condition the column. To verify the performance of the system, chromatograph the Resolution Solution, and record the peak responses as directed under Procedure. The relative retention times are about 0.8 for valerophenone, and 1.0 for ibuprofen. The resolution, R, between the valerophenone peak and the ibuprofen peak is not less than 2.0.

The assay procedure consists of injecting about 5 μL of the Test Preparation into the chromatograph, recording the chromatogram, and measuring the peak responses.

The percentage of each impurity (%IMP$_i$) is calculated using:

$$\%IMP_i \quad = \quad \{ R_i / R_t \} \; 100$$

where R_i is the response of an individual peak (other than the solvent peak and the main ibuprofen peak), and R_t is the sum of the responses of all the peaks (excluding that of the solvent peak).

The specification is that not more than 0.3% of any individual impurity is found, and the sum of all the individual impurities found does not exceed 1.0%.

4.1.7 Limit of 4-*Iso*butylacetophenone

Using the chromatograms of the Assay preparation and the 4-*iso*butyl-acetophenone standard solution obtained as directed in the Assay, calculate the percentage of 4-*iso*butylacetophenone (4-IBAP) in the test article taken using:

$$\%(4\text{-}IBAP) \quad = \quad \{ (C/W) \, (R_U / R_S) \} \; 10000$$

where C is the concentration (in mg/mL) of 4-*iso*butylacetophenone in the 4-*iso*butylacetophenone standard solution, W is the weight (in mg) of test article taken to make the Assay preparation, and R_U and R_S are the ratios of the 4-*iso*butylacetophenone peak response to the valerophenone peak response obtained from the Assay preparation and the Standard preparation, respectively. The specification is that not more than 0.1% is found.

4.1.8 Assay

The assay value of the test article is established using high-performance liquid chromatography. The mobile phase is prepared by dissolving 4.0 g of chloroacetic acid in 400 mL of water, and adjusting with ammonium hydroxide to a pH of 3.0. To this, one adds 600 mL of acetonitrile, and the solution is filtered and degassed. The Internal standard solution consists of a solution of valerophenone in Mobile phase having a concentration of about 0.35 mg/mL. The Standard preparation is made by dissolving an accurately weighed quantity of Ibuprofen reference standard in the Internal standard solution, to obtain a solution having a known concentration of about 12 mg/mL. A 4-*iso*butylacetophenone standard solution is prepared by quantitatively dissolving an accurately weighed quantity of 4-*iso*butylacetophenone in acetonitrile to obtain a solution having a known concentration of about 0.6 mg/mL. Add 2.0 mL of this stock solution to 100.0 mL of Internal standard solution, and mix to obtain a solution having a known concentration of about 0.012 mg of 4-*iso*butylacetophenone per milliliter.

The Assay preparation is made by transferring about 1200 mg of the test article (accurately weighed) to a 100-mL volumetric flask, diluting to volume with Internal standard solution, and mixing.

The liquid chromatograph is equipped with a 254-nm detector and a 4.6-mm x 25-cm column that contains packing L1. The flow rate is about 2 mL/minute. To verify the performance of the system, chromatograph the Standard preparation, and record the peak responses as directed under Procedure. The resolution, R, between the ibuprofen and internal standard peaks is not less than 2.5, and the relative standard deviation for replicate injections is not more than 2.0%. Then chromatograph the 4-*iso*butylacetophenone standard solution, and record the peak responses as directed under Procedure. The relative retention times are about 1.0 for valerophenone and 1.2 for 4-*iso*butylacetophenone, and the tailing factors for the individual peaks are not more than 2.5. Furthermore, the resolution between the valerophenone peak and the 4-*iso*butylacetophenone peak is not less than 2.5, and the relative standard deviation for replicate injections is not more than 2.0%.

To begin the Assay Procedure, separately inject equal volumes (about 5 μL) of the Standard preparation, the Assay preparation, and the 4-*iso*butyl-acetophenone standard solution into the chromatograph, record the chromatograms, and measure the responses for the major peaks. The relative retention times are about 1.4 for the internal standard and 1.0 for ibuprofen.

Calculate the quantity (in mg) of analyte in the portion of Ibuprofen taken using the formula:

$$\text{mg IBU} \quad = \quad \{ \, C \, (R_U / R_S) \, \} \; 100$$

where C is the concentration (in mg/mL) of Ibuprofen standard in the Standard preparation, and R_U and R_S are the peak response ratios obtained from the Assay preparation and the Standard preparation, respectively.

The specification is that USP Ibuprofen contains not less than 97.0 % and not more than 103.0 % of $C_{13}H_{18}O_2$, calculated on the anhydrous basis.

4.2   Elemental Analysis

The calculated elemental percentages for ibuprofen are as follows:

Carbon:	75.69%
Hydrogen:	8.80%
Oxygen:	15.51%

4.3   High Performance Liquid Chromatographic Methods of Analysis

Several chromatographic systems have been employed for the analysis of *rac*-ibuprofen in various pharmaceutical dosage forms. While the most common methods employ typical reverse phase HPLC [32-37], special reverse phase HPLC methods have been developed for unique cases. Examples include an LC/MS method [38] which requires a volatile mobile phase, a method for ointments [39] that employs column washing and a reverse phase ion-pairing HPLC [40] method that uses modified silica gel.

An example method which is appropriate for *rac*-ibuprofen tablets and caplets employs injection of a 0.5 mg/mL sample onto a 150 x 4.6 mm Phenomenex C_8 reverse phase HPLC column [23]. The method uses 55:45 acetonitrile / 0.1 M acetic acid as the mobile phase, eluted at a flow rate of 1.5 mL/min. Detection is on the basis of the UV absorbance at 254 nm. The retention time for *rac*-ibuprofen is approximately 5 minutes.

An effective method employed for the analysis of *rac*-ibuprofen suspensions uses a 15 cm Waters Spherisorb S5C$_8$ column reverse phase HPLC column [23]. The mobile phase is 44:56 tetrahydrofuran / 20 mM monobasic potassium phosphate (pH 2.0), and the flow rate is 1.0 mL/min. Due to the use of THF in the mobile phase, PEEK tubing cannot be used under these HPLC conditions. Detection is on the basis of the UV absorbance at 254 nm, and the retention time for *rac*-ibuprofen is approximately 5 minutes.

4.4   Gas Chromatographic Methods of Analysis

Gas chromatographic methods can be employed as methods of analysis, but this approach often requires derivatization of the analyte before injection [41-43].

4.5   Thin-Layer Chromatographic Methods of Analysis

The British Pharmacopoeia specifies a TLC method to be used during the Identification sequence of testing [44], which follows the outlines of general method (2.2.27) and uses silica gel H as the coating substance. The test solution is prepared by dissolving 50 mg of the substance to be examined in methylene chloride, and diluting to 10 mL with the same solvent. The reference solution is prepared by dissolving 50 mg of ibuprofen reference standard in methylene chloride, and diluting to 10 mL with the same solvent.

The method is initiated by separately applying 5 µL of each solution to the plate. The place is developed over a path of 10 cm, using a mixture of 5:25:75 v/v/v anhydrous acetic acid / ethyl acetate / hexane. The plate is allowed to dry at 120°C for 30 minutes. After that, the plate is lightly sprayed with a 10 g/L solution of potassium permanganate in dilute sulfuric acid, and heated at 120°C for 20 minutes. At the conclusion of the method, the plate is examined under 365 nm ultraviolet light.

4.6   Titrimetric Analysis

The British Pharmacopoeia also specifies a titration method for the determination of the assay value of ibuprofen [44]. To perform the method, one dissolves 0.450 g of the test article in 50 mL of methanol, and add to this 0.4 mL of phenolphthalein solution R1. This solution is titrated with 0.1 M NaOH volumetric standard solution until a red color is obtained. Correction of the titrant volume is required through he performance of a blank titration. Each milliliter of 0.1 M NaOH VS is equivalent to 20.63 mg of $C_{13}H_{18}O_2$.

4.7   Determination in Body Fluids and Tissues

HPLC [45-47] methods represent the most commonly used approach for the analysis of ibuprofen and metabolites compounds in biological fluids, and the general topic has been reviewed [48].

A typical sample preparation for HPLC analysis consists of the addition of 50 μL of plasma to 150 μL of 95% acetonitrile, containing 25 mg/L of internal standard. The mixture is vortexed for 30 seconds, centrifuged for 1 minute through a microfuge filter, and the supernatant is injected onto the HPLC [46].

Several GC methods using derivatized ibuprofen also have been published [49,50].

5.   Stability

5.1   Solid-State Stability

In the solid-state, *rac*-ibuprofen is considerably stable when subjected to both ambient and accelerated stability testing. Less than 0.1% degradation of *rac*-ibuprofen is observed upon exposure over several months to all of the following environmental conditions over several months [51]:

a.   Ambient temperature and humidity.

b.   Ambient temperature, 100% relative humidity.

c.   37°C and 60°C, ambient relative humidity.

d.   37°C, 100% relative humidity.

e.   UV light, ambient temperature.

5.2   Solution-Phase Stability

In the solution phase, *rac*-ibuprofen has been shown to be relatively stable, even when exposed to harsh conditions such as 1 N NaOH, 1 N HCl, or 50% H_2O_2 [23]. Degradants that form in quantities less than 0.1 % include *iso*butylacetophenone (IBAP) and 2-(4-isobutyrylphenyl)-propionic acid:

2-(4-isobutyrylphenyl)-proprionic acid   IB AP

IBAP has been proposed [52] to form *via* radical induced decarboxylation, followed by benzylic oxidation. Suitable HPLC methods for separating *rac*-ibuprofen from its degradants include the USP [31] ibuprofen assay method and the tablet and caplet assay methods [32-37].

6.   Drug Metabolism and Pharmacokinetics

6.1   Pharmacokinetics

Many studies on the pharmacokinetics and metabolism of *rac*-ibuprofen
and its individual enantiomers have been reported [1,3,53,54]. After oral
administration of *rac*-ibuprofen, well over 80% of the dose is absorbed [3]
from the GI tract (mainly the intestine [55]) and peak plasma levels (t_{max})
are attained within 1.5 - 2 hours. While variations in the extent of
absorption and overall pharmacokinetics of *rac*-ibuprofen have been
observed depending on the formulation, ingestion of food has been shown
to have little effect [1].

Once in serum, greater than 98% of *rac*-ibuprofen is bound to plasma
proteins such as albumin [56]. The apparent half life has been observed to
be between approximately 2-3 hours [1]. Detailed reports on the rate of
ibuprofen absorption [1] and the effects of stereochemistry on
pharmacokinetics and efficacy [54] are reported elsewhere.

6.2   Metabolism

There are 3 routes of metabolism [3] of *rac*-ibuprofen, consisting of
oxidation, acylglucuronide conjugation, and stereochemical inversion of
the *R*-(-)-enantiomer to the *S*-(+) enantiomer. These are illustrated in
Figure 11.

Oxidation is the major metabolic pathway for ibuprofen, and in humans
more than 60% of an oral dose is recovered in the urine as 2-hydroxy-
ibuprofen and 3-carboxy-ibuprofen and their glucuronide conjugates
[57,58]. Regarding the oxidative pathway, there is considerable evidence
that *rac*-ibuprofen hydroxylation is mediated by the cytochrome P450
family of enzymes [59]. The second metabolic pathway, direct
conjugation of *rac*-ibuprofen to glucuronic acid, is a relatively minor
pathway in humans, contributing to less than 10% of an oral dose [60].

Many groups have studied the *in vivo* process by which (*R*)-ibuprofen
undergoes stereochemical inversion. The interest began when studies in
the early 1970's revealed that most urinary metabolites of ibuprofen were
of the (*S*)-configuration, regardless of which enantiomer was initially
administered [61]. More recently, it was observed that more than 60% of
an oral dose of (*R*)-ibuprofen is converted to the (*S*)-enantiomer in humans

Figure 11.   Metabolism of ibuprofen.

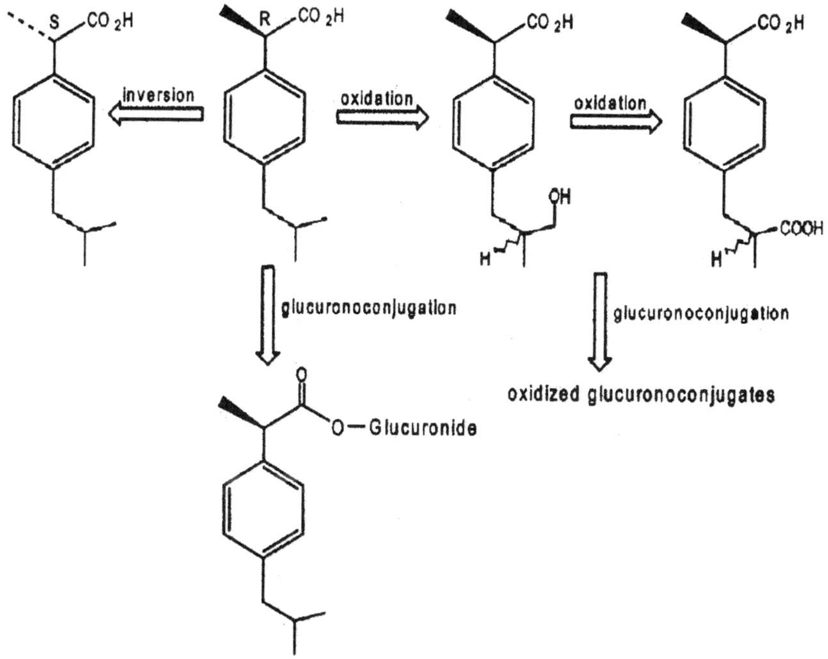

[62]. It has been shown that the mechanism of metabolic chiral inversion of (*R*)-ibuprofen involves enzyme catalyzed enantioselective thioesterification of the carboxylate with CoA, followed by epimerization of the subsequent thioester. Enzymatic hydrolysis of the thioester affords the inverted (*S*)-enantiomer of ibuprofen [63].

References

1.   (a) An excellent review of many of the chemical, physiologic and clinical aspects of ibuprofen is found in ***Ibuprofen***, K.D. Rainsford, ed., Taylor and Francis Publishers, London, UK, 1999.

2.   The clinical safety and efficacy of ibuprofen has been reviewed extensively; for example see G.L. Royer, B.S. Seckman, and I.R. Welshman, *Am. J. Med.*, **July 13 issue**, 25 (1984).

3.   J.M. Mayerand and B. Testa, *Drugs of the Future*, **22**, 1347 (1997).

4   Chapter 27, in ***The Pharmacological Basis of Therapeutics***, J.G. Hardman, L.E. Limbrid, P.B. Molinoff, R.W. Ruddon, and A.G. Gilman, eds., McGraw Hill, NY, 1996.

5.   G.P. Stahly and R.M. Starrett, in ***Chirality in Industry***, A.N. Collins, ed., 1997, pp. 19-38.

6   R.A. Dytham, **GB Patent 1160725**, 1969.

7.   V. Elango, M.A. Murphy, B.L. Smith, K.G. Davenport, G.N. Moft, and G.L. Moss, **Eur. Pat. Appl. 284,310**, 1988

8.   J.F. McConnell, *Cryst. Struct. Comm.*, **3**, 73 (1974).

9.   N. Shankland, C.C. Wilson, A.J. Florence, and P.J. Cox, *Acta Cryst.*, **C53**, 951 (1997).

10. N. Shankland, A.J. Florence, P.J. Cox, C.C. Wilson, and K. Shankland, *Int. J. Pharm.*, **165**, 107 (1998).

11. A.K. Romero, L. Savastano, and C.T. Rhodes, *Int. J. Pharm.*, **99**, 125 (1993).

12. V. Labhasetwar, S.V. Deshmukh, and A.K. Dorle, *Drug Dev. Indust. Pharm.*, **19**, 631 (1993).

13. G.M. Khan and Z. Jiabi, *Drug Dev. Indust. Pharm.*, **24**, 463 (1998).

14. J.A. McCauley, *AIChE Symp. Series*, **87**, 58 (1991).

15. A.J. Romero, G. Lukas, and C.T. Rhodes, *Pharm. Acta. Helv.*, **66**, 34 (1991).

16. ICDD pattern 32-1723, grant-in-aid, (1981) primary reference, P. Gong, Polytechnic Inst. of Brooklyn, NY, USA.

17. G.P. Stahly, A.T. McKenzie, M.C. Andres, C.A. Russell, S.R. Byrn, and P. Johnson, *J. Pharm. Sci.*, **86**, 970 (1997).

18. H.H. Tung, S. Waterson, S. Reynolds, and E. Paul, *AIChE Symp. Series*, **87**, 64 (1991).

19. T. Manimaran and G.P. Stahly, *Tetrahedron Asymmetry*, **4**, 1949 (1993).

20. E.J. Ebbers, B.J.M. Plum, G.J.A. Ariaans, B. Kaptein, Q.B. Broxterman, A. Bruggink, and B. Zwanenburg, *Tetrahedron Asymmetry*, **8**, 4047 (1997).

21. S.K. Dwivedi, S. Sattari, F. Jamali, and A.G. Mitchell, *Int. J. Pharm.*, **87**, 95 (1992).

22. **The Merck Index**, 10th edn., 1983, p. 712.

23. Unpublished results, McNeil Consumer Healthcare, Fort Washington, PA.

24.   K.D. Ertel, R.A. Heasley, C. Koegel, A. Chakrabarti, and J.T. Carstensen, *J. Pharm. Sci.*, **79**, 552 (1990).

25.   S. Schmid, C.C. Mueller-Goymann and P.C. Schmitd, *Int. J. Pharm.*, **197**, 35 (2000).

26.   Since ibuprofen is only sparingly soluble in water, pKa values in pure water have been calculated from values determined in binary water/organic solvent mixtures. See for example (a) K. Takas-Novak, K.J. Box, and A. Avdeef, *Int. J. Pharm.*, **151**, 235 (1997) or (b) C. Rafols, M. Roses, and E. Bosch, *Anal. Chim. Acta*, **350**, 255 (1997).

27.   H.G. Brittain, unpublished results obtained using ACD/PhysChem, Advanced Chemistry Development, Toronto, CA.

28   Solubility data provided by BASF Corporation, Bishop, TX.

29.   G. Lelsing, R. Resel, F. Stelzer, S. Tasch, A. Lanzier, and G. Hantich, *J. Clin. Pharmacol.*, **36**, 3S (1996).

30.   ***Hewlet Packard MS Chemstation Library***, John Wiley & Sons, Inc., New York, 1995.

31.   ***USP 24***, The United States Pharmacopoeial Convention, Rockville, MD, p. 854.

32.   J. Ravi, C.L. Jain, and Y.K. Bansal, *Indian Drugs*, **27**, 615 (1990).

33.   A.K. Sen, A. Bandyopadhyay, G. Podder, and B.J. Chowdhury, *J. Indian Chem. Soc.*, **67**, 443 (1990).

34.   B.M. Lampert and J.T. Stewart, *J. Chromatogr.*, **504**, 381 (1990).

35.   S. Husain, A.S.R. Murty, and R. Narasimha, *Indian Drugs*, **26**, 557 (1989).

36.   J.C. Tsao and T.S. Savage, *Drug Dev. Ind. Pharm.*, **11**,1123 (1985).

37.   M.K. Aravind, J.N. Miceli, and R.E. Kauffman, *J. Chromatogr.*, , **308**, 350 (1984).

38. K. Kamei, K. Kitahara, A. Momose, K. Matsuura, and H. Yuki, *Shitsutyo Bunseki*, **36**, 115 (1988).

39. V.E. Haikala, I.K. Heimonen, and H.J. Vuorela, *J. Pharm. Sci.*, **80**, 456 (1991).

40. G.R. Rao, A.B. Avadhanulu, and A.R.R. Pantulu, *East. Pharm.*, **34**, 119 (1991).

41. S.K. Pant and C.L. Jain, *Indian Drugs*, **28**, 262 (1991).

42. L Heikkinen, *Acta Pharm. Fenn.*, **92**, 275 (1983).

43. Z. Budvari-Barany, G. Radeczky, A. Shalaby, and G. Szasz, *Acta Pharm. Hung.*, **59**, 49 (1989).

44. *The British Pharmacopoeia 1999*, the Stationary Office, 1999.

45. A.M. Rustum, *J. Chromatogr. Sci.*, **29**, 16 (1991).

46. A. Shah and D. Jung, *Chromatographia*, **378**, 232 (1986).

47. N. Rifia, M. Sakamoto, T. Law, V. Gaipchian, N. Harris, and A. Colin, *Clin. Chem.*, **42**, 1812 (1996).

48. A. Hercegova and J. Polonsky, *Pharmazie*, **54**, 479 (1999).

49. D.G. Kaiser and J. Vangiessen, *J. Pharm. Sci.*, **63**, 219 (1974).

50. D.G. Kaiser and R.S. Martin, *J. Pharm. Sci.*, **67**, 627 (1978).

51. *Ibuprofen USP Customer Dossier*, BASF Corporation, Bishop, TX.

52. S. Chauhan, and B. Sahoo, *Bioorg. Med. Chem.*, **7**, 2629 (1999).

53. K.S. Albert, *Am. J. Med.*, **July 13**, 40 (1984).

54. G.F. Lockwood, K.S. Albert, M.S. Gillespie, M.D. Bole, M.D. Harkcom, B.S. Szpunar, and J.G. Wager, *Pharmacol Therap.*, **34**, 97 (1983).

55.   S.S. Adams, G. Gough, E. Cliffe, B. Lessel, and R. Mills, *Tox. Appl. Pharm.*, **15**, 310 (1969).

56.   A.M. Evans, *J. Clin. Pharmacol.*, **36**, 7 (1996).

57.   A.C. Rudy, K.S. Anlikerand, and S.D. Hall, *J. Chromatog.*, **538**, 395 (1990).

58.   A.C. Rudy, P.M. Knight, D.C. Brater, and S.D. Hall, *J. Pharmacol. Exp. Therap.*, **273**, 8893 (1995).

59.   T.D. Leeman, *Drug Exp. Clin. Res.*, **19**, 189 (1993).

60.   E.J.D. Lee, K. Williams, R. Day, G. Graham, and D. Champion, *J. Clin. Pharmacol.*, **19**, 669 (1985).

61.   R.F.N. Mills, S. Adams, E. Cliffe, W. Dickenson, and J. Nicholson, *Xeneobiotica*, **3**, 589 (1973).

62.   H. Cheng, J. Rogers, J. Demetriades, S. Holland, J. Seibold, and E. Depuy, *Pharm. Res.*, **11**, 824 (1994).

63.   T.A. Baellie, W. Adams, and D. Kaiser, *Pharmacol. Exp. Therap.*, **249**, 517 (1989).

ONDANSETRON HYDROCHLORIDE

Isam Ismail Salem,[1] José M. Ramos López [2],
and Antonio Cerezo Galán,[1]

(1) Department of Pharmacy and Pharmaceutical Technology
Faculty of Pharmacy
University of Granada
18071-Granada
Spain

(2) Scientific Instrumentation Center
University of Granada
18071-Granada
Spain

Contents

1.   Description

1.1   Nomenclature

1.1.1   Chemical Name

1,2,3,9-Tetrahydro-9-methyl-3-[(2-methyl-1H-imidazol-1-yl)methyl]-4H-carbazol-4-one

(±)–1,2,3,9-Tetrahydro-9-methyl-3-[(2-methyl-1H-imidazol-1-yl)methyl]-4H-carbazol-4-one, monohydrochloride, dihydrate

1.1.2   Nonproprietary Names

Ondansetron; GR-38032

Ondansetron hydrochloride (BANM, USAN, rINNM)

1.1.3   Proprietary Names

Zofran (GlaxoWellcome)

1.2   Formulae

1.2.1   Empirical

Ondansetron: $C_{18}H_{19}N_3O$

Ondansetron hydrochloride dihydrate: $C_{18}H_{19}N_3O \cdot HCl \cdot 2H_2O$

1.2.2   Structural

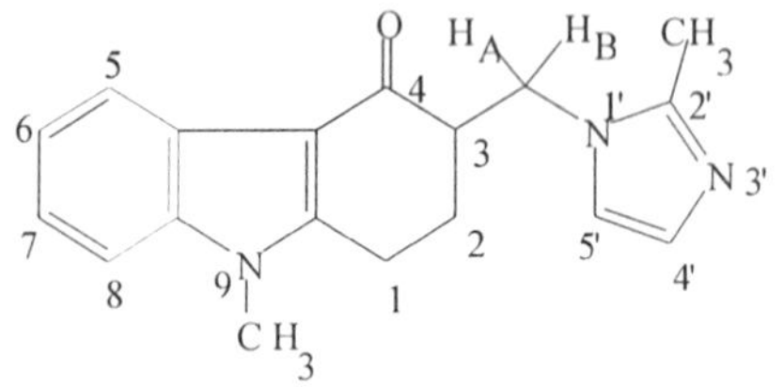

1.3   Molecular Weight and CAS Number

Ondansetron:	293.37	99614-02-5
Ondansetron hydrochloride dihydrate:	365.87	103639-04-9

1.5   Appearance

Ondansetron hydrochloride is obtained as a white or off-white powder.

1.6   Uses and Applications

Ondansetron hydrochloride is the soluble form of ondansetron, a tetrahydrocarbazolone derivative with an imidazolylmethyyl group. It is the first of a class of selective serotonin 5-HT3 receptor antagonists indicated for the prevention of nausea and vomiting associated with initial and repeat courses of emetogenic cancer chemotherapy, radiotherapy or anesthesia, and surgery [1]. There are several routes of administration and dosages of ondansetron hydrochloride, such as sterile injection for intravenous (I.V.) or intramuscular (I.M.) administration, tablets, and oral solutions. FDA approval for ondansetron (under the trade name of Zofran) was obtained by Glaxo-Wellcome on January 1991.

Ondansetron was proposed to be investigated as a possible treatment of opioid withdrawal signs in heroin addicts [2]. Studies in animals had shown efficacy in decreasing the intensity of withdrawal signs in rats, such as increased defecation, jumping, and wet-dog shakes. It also elevated the nociceptive threshold values that were decreased by precipitated withdrawal.

2.   Methods of Preparation

The original synthetic routes were first described by Coates et al. [3,4]. The preparation of ondansetron was achieved by quarternizing (aminomethyl)carbazolone with methyl iodide, and reacting the product with 2-methylimidazole. After that, N-alkalizing by (bromomethyl)-cyclopropane gave carbazolone as in Scheme 1. Coates et al. proposed a preparation method for ondansetron that is illustrated in Scheme 2.

Nevertheless, most of the processes here mentioned and other published methods of synthesis for ondansetron [5,6] have drawbacks, such as lengthy steps and the use of dangerous reagents. In order to overcome such inconveniences, Kim et al. have proposed in 1997 a new two-steps synthetic route for ondansetron with a 70% overall yield, amendable to large-scale industrial synthesis [7]. The process was accomplished from readily available N-methyltetrahydrocarbazolone employing direct Mannich α-methylenation and alumina catalyzed Michael addition of imidazole to exocyclic α, β-unsaturated ketone as the key steps. This is illustrated in Scheme 3.

Scheme 1.   Principal synthetic routes to ondansetron [3].

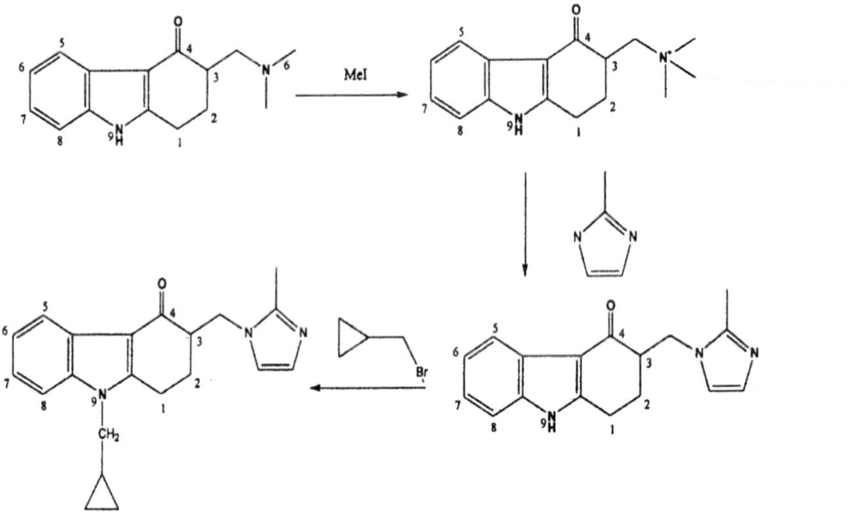

Scheme 2.   Preparation method for ondansetron [4].

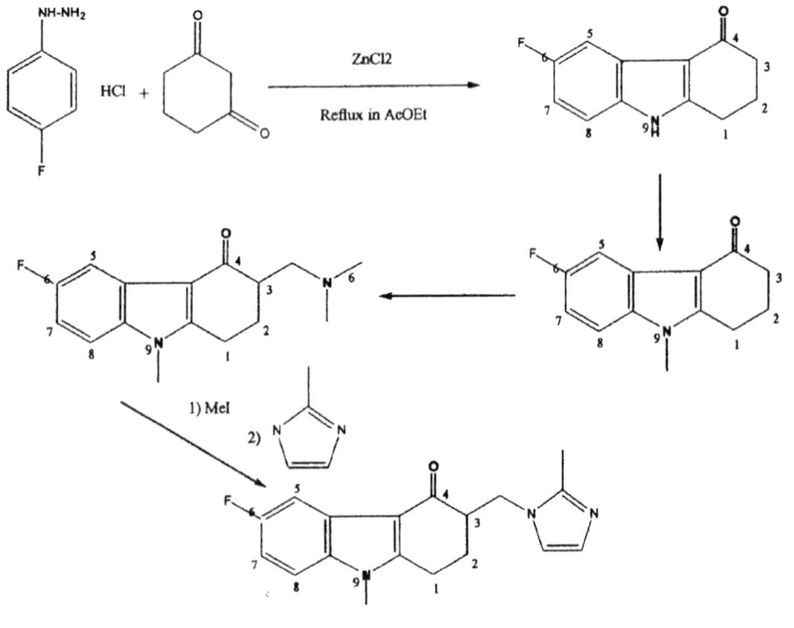

Scheme 3.   Alternate synthesis of ondansetron [7].

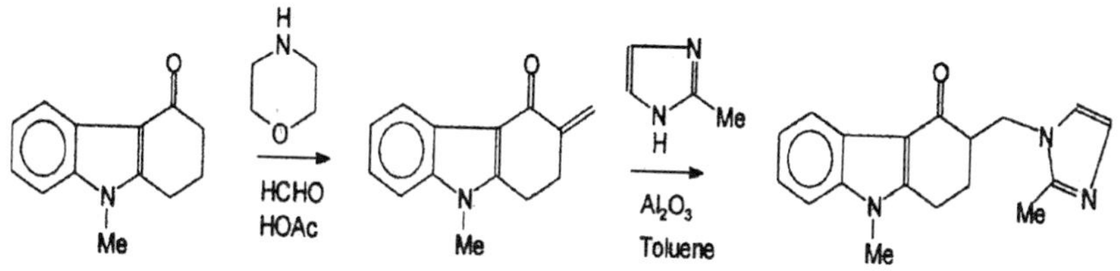

3.   Physical Properties

3.1   Crystallographic Properties

3.1.1   Single Crystal Structure

Mohan and Ravikumar concluded that ondansetron hydrochloride dihydrate crystallized in the monoclinic crystal system, and that the space group was P2$_1$/c [8]. The size of the unit cell was characterized by the dimensions of a = 15.185(2) Å, b = 9.810(1) Å, and c = 12.823(2) Å. Additional characteristics of the unit cell were the monoclinic angle (β) of 100.89(2)°, and the unit cell contained 4 molecules. The calculated density was 1.295, with R= 0.051 and Rw = 0.062 being determined for 1918 reflections. In this work, it was concluded that the methyl substituted imidazole ring is approximately perpendicular to the carbazole plane with a dihedral angle of 87.0(1)°.

3.1.2   X-Ray Powder Diffraction Pattern

The x-ray powder diffraction pattern of ondansetron hydrochloride was obtained on a Philips diffractometer system (model PW1710). The pattern, shown in Figure 1, was obtained using nickel filtered copper radiation (Cu anode 2000W, λ = 1.5405 Å). A full data summary is provided in Table 1.

3.2   Thermal Methods of analysis

3.2.1   Melting Behavior

The literature melting point reported for the free base of ondansetron (crystallized from methanol) is 223–232°C [7,9]. For ondansetron hydrochloride dihydrate that had been crystallized from aqueous isopropanol, the melting range was reported to be 178.5–179.5°C [9].

Figure 1.   X-ray powder diffraction pattern of ondansetron
             hydrochloride dihydrate.

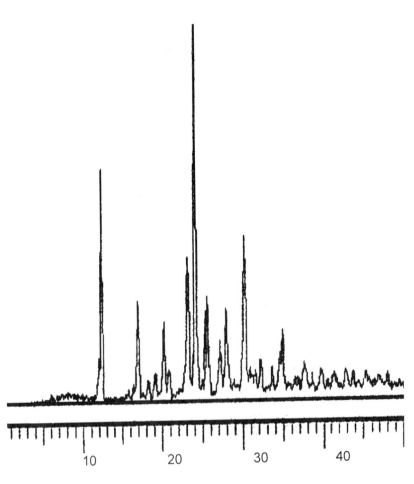

Scattering Angle (degrees 2-θ)

Table 1

Crystallographic Data Associated with the X-ray Powder Pattern of
Ondansetron Hydrochloride

Scattering Angle (degrees 2-θ)	d-spacing (Å)	Relative Intensity (%)
11.96	7.39	8.59
12.28	7.20	55.10
16.84	5.26	22.51
17.08	5.19	6.44
18.04	4.91	3.47
20.20	4.39	16.75
23.08	4.26	6.15
23.96	3.85	31.28
24.36	3.71	100.00
25.48	3.65	6.33
25.72	3.49	22.44
27.08	3.46	11.65
27.88	3.29	11.39
30.12	3.20	18.20
30.84	2.96	35.58
31.56	2.89	3.32
32.20	2.83	3.35
33.56	2.77	7.71
34.52	2.66	5.26
34.92	2.60	7.99
35.08	2.56	14.21
37.40	2.60	3.23
37.56	2.40	4.00
39.56	2.39	4.41
42.68	2.27	4.53
43.72	2.11	4.84
45.08	2.06	3.65
47.96	2.00	3.77
47.96	1.89	3.29
77.96	1.22	4.35

3.2.2   Differential Scanning Calorimetry

The thermal behavior of ondansetron hydrochloride was examined by DSC, using a Shimazu DSC-50 differential scanning calorimeter. The system was calibrated with a high purity sample of Indium. Ondansetron hydrochloride samples of 5-6 mg were scanned at 5°C/min over a temperature range of 30 to 400°C. Peak transition and enthalpies of fusion were determined for all samples.

The DSC thermogram of ondansetron hydrochloride dihydrate features a single sharp melting endotherm, having a peak temperature of 183°C. Onset and endset temperatures were 178.5 and 192.7°C, respectively. Integration of the melting endotherm permitted an estimation of the enthalpy of fusion (ΔH) for ondansetron hydrochloride as 80.9 J/g. The large endotherm observed at 106°C ($\Delta H = 182$ J/g) corresponds to the dehydration process. The exothermic decomposition peak has an onset temperature at 313°C, and a peak temperature at 331°C.

Unexpectedly, differences have been found in the DSC behavior of two different batches of ondansetron hydrochloride used to produce tablets. DSC curves of these two batches are shown in Figures 3 and 4. In the second tested batch of this active, other thermal events besides the characteristic melting endotherm of ondansetron (183 °C) were noted at 208.5 and 225.5°C with a total $\Delta H = 40.32$ J/g (Figure 4). In addition, a 6.8°C shift of the ondansetron peak temperature was observed with a mean $\Delta H = 37.27$ J/g. The origin of these variations is not yet understood.

3.2.3   Thermogravimetric Analysis

Ondansetron hydrochloride TGA thermograms were recorded on a Shimazu TGA 50H thermogravimetric analyzer, that was on-line with a Fisons Instruments Thermolab mass detector and a Nicolet TGA Magna-IR 550 interface. The instrument was calibrated using the latent heat of fusion of indium. The experiments were carried out in flowing nitrogen or air (100 mL/min) at different heating rates (from 10°C/min to 20°C/min). The samples were analyzed over a temperature range between 20°C and 400°C, and gases that were produced during the experiments were submitted for mass and infrared analysis.

Figure 5 shows the thermogravimetric analysis of ondansetron hydrochloride obtained in an aluminum closed pan. TGA studies revealed two different regions, the first of these was complete by approximately

Figure 2.   DSC thermogram of ondansetron hydrochloride dihydrate.

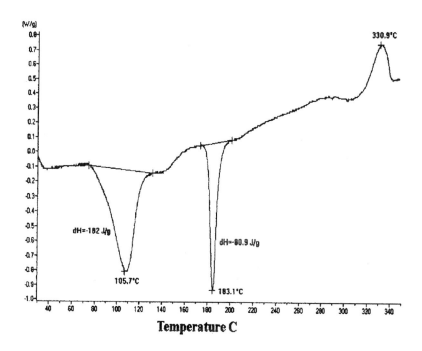

Figure 4.   DSC curve of the second tested batch of ondansetron
hydrochloride dihydrate.

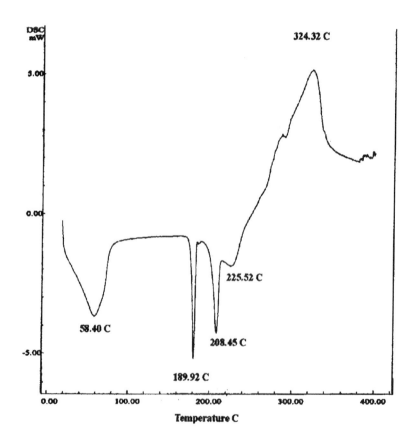

Figure 5. Thermogravimetric analysis of ondansetron hydrochloride, and the first derivative of the thermogram.

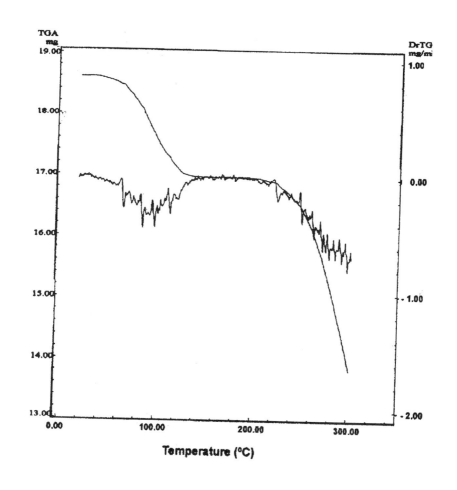

Figure 6. Gases generated during the thermogravimetric analysis of ondansetron hydrochloride.

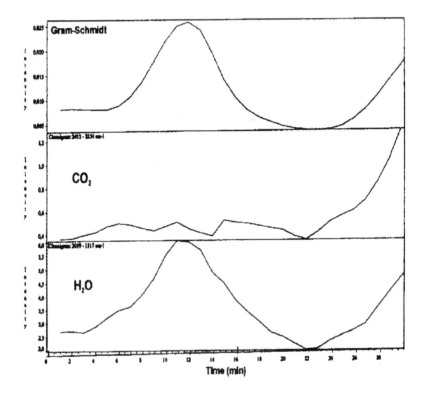

135°C, and reflects the loss of lattice water. The magnitude of this weight loss is about 9%, which is consistent with the anticipated water content of the monohydrate phase. The second weight loss begins at 205 °C, and the IR and mass spectral analysis of the generated gases (Figure 6) revealed oxidative decomposition of ondansetron above this temperature.

3.3   Solubility Characteristics

Ondansetron is a weak base (pKa = 7.4), and under the acidic conditions is water-soluble. The natural pH of ondansetron hydrochloride solutions is about 4.5 to 4.6 [10,11]. The solubility is markedly reduced in solutions for which the pH greater than or equal to 6. Precipitation of ondansetron (as the free base) occurs in solutions with a pH of 5.7 or more [12]. Re-dissolution of the ondansetron precipitate occurs at pH 6.2 when titrated with hydrochloric acid, and precipitation by combination with alkaline drugs has been observed [13].

3.4   Spectroscopy

3.4.1 UV/VIS Spectroscopy

The ultraviolet absorption spectrum of a 1% solution of ondansetron hydrochloride was obtained using a Perkin Elmer Lambda 2 UV visible spectrophotometer and 1 cm quartz cells, over a wavelength range of 200 to 400 nm. The spectrum for substance dissolved in acidified water (0.1 N HCl) is shown in Figure 7, and exhibits maxima at: 210, 249, 266, and 310 nm. When dissolved in methanol, the UV absorption spectrum (Figure 8) exhibits maxima at: 212, 246, 266 and 303 nm. These slight shifts in absorption maxima indicate the presence of an aromatic chromophore in the substance.

3.4.2 Vibrational Spectroscopy

The infrared absorbance spectrum of ondansetron hydrochloride was recorded using a Nicolet 20SXB FT-IR spectrophotometer over the range of 400 to 4000 cm^{-1} at a resolution of 2 cm^{-1}. The spectrum obtained in a KBr pellet is shown in Figure 9, and structural assignments for some of the characteristic absorption bands are listed in Table 2.

The infrared spectrum of the second lot of ondansetron hydrochloride tested is located in Figure 10.

Figure 7.   Ultraviolet absorption spectrum of ondansetron
             hydrochloride dissolved in acidified water.

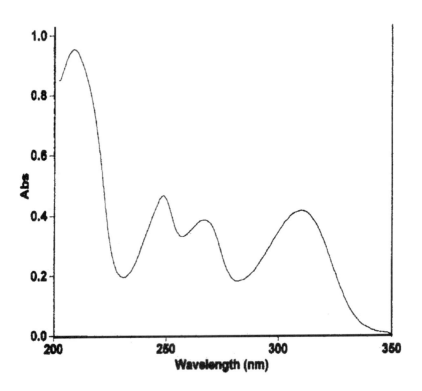

Figure 8.   Ultraviolet absorption spectrum of ondansetron
hydrochloride dissolved in acidified methanol.

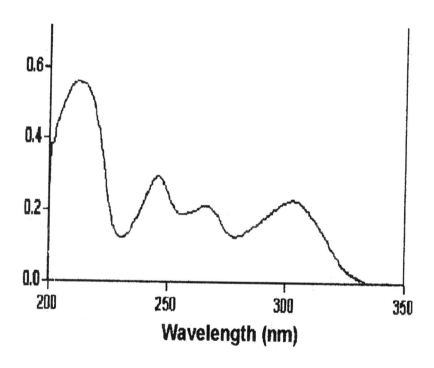

Figure 9. Infrared absorption spectrum of ondansetron hydrochloride in a KBr pellet.

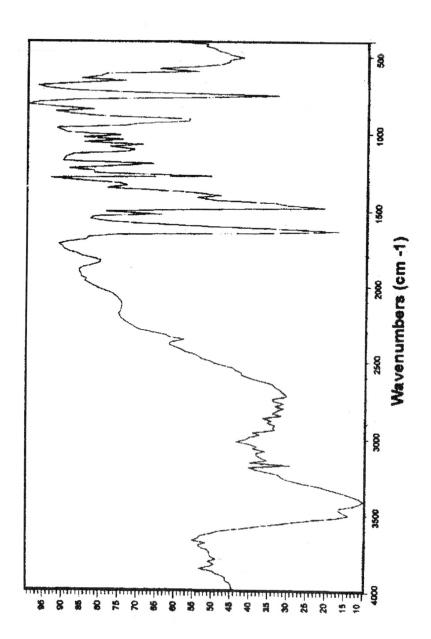

Table 2

Assignments for the Principal Infrared Absorption Bands of
Ondansetron Hydrochloride

Peak Maximum (cm^{-1})	Assignment
756	*o*-disubstituted benzene
1279	C-N
1458 and 1479	CH3
1531	C=C aromatic
1638	C=N, C=O in six member ring
3410	H_2O

320   I.I. SALEM, K.J.M.R. LÓPEZ, AND A.C. GALÁN

Figure 10.   Infrared absorption spectrum of the second lot of
ondansetron hydrochloride (KBr pellet).

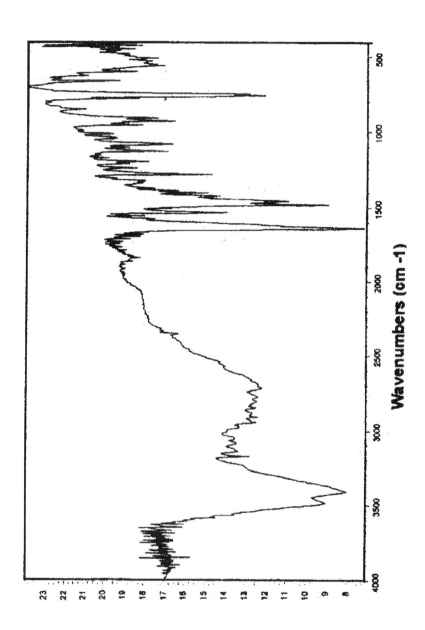

3.4.3   Nuclear Magnetic Resonance Spectrometry

3.4.3.1 ^{1}H-NMR Spectrum

The ^{1}H-NMR spectrum for ondansetron hydrochloride dissolved in d_6-DMSO (Figure 13) was obtained at a 300 MHz frequency using Bruker AM 300, tetramethylsilane TMS was used as the internal standard. The one-dimensional spectrum is shown in Figure 11, along with a proton numbering assignment scheme. Assignments of the resonance bands observed in the ^{1}H-NMR spectrum are found in Table 3.

3.4.3.2 ^{13}C-NMR Spectrum

The ^{13}C-NMR spectrum of ondansetron hydrochloride was obtained at 75 MHz, using the same instrumental system as described in the ^{1}H-NMR section. Tetramethylsilane was used as the internal standard. The one-dimensional spectrum is shown in Figure 12, together with a figure illustrating the resonance assignments.

3.4.4   Mass Spectrometry

The fast atom bombardment (FAB) high-resolution mass spectrum of ondansetron hydrochloride was recorded using an EBE Autospec-Q (Micromass Instruments, UK). The LSIMS (FAB) technique was applied to samples of ondansetron hydrochloride using thioglycol with 1% NaI as a matrix and a 25 kV acceleration voltage.

As shown in Figure 13, the high resolution mass spectrum showed a $[M+1]^+$ peak at m/z 294.160608 Da, which analyzes as follows:

Ion	Formula	Theoretical Mass	Empirical Mass	Error (ppm)	DBE
$[M+1]^+$	$C_{18}H_{20}N_3O$	294.160637	294.160608	0.1	10.5

A low-resolution mass spectrum was obtained by electron impact (EI) at 70 eV using Platform II (Micromass Instruments, UK). The EI mass spectrum showed peaks at the molecular ion values of m/z = 293 (100%). Other major peaks were detected at m/z = 211 (64%), m/z = 198 (84%), m/z = 143 (66%), m/z = 115 (44%), and m/z = 96 (95%).

Figure 11. One-dimensional ^{1}H-NMR spectrum of ondansetron
           hydrochloride.

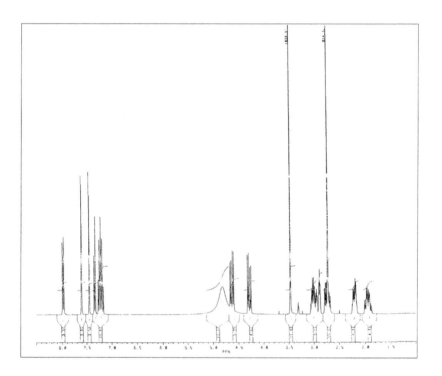

Table 3

Assignment of the Resonance Bands Observed in the
^{1}H-NMR Spectrum of Ondansetron Hydrochloride

Assignment	Chemical Shift (ppm)	Multiplicity	Coupling Constant (Hz)
$CH_3 - C_{2'}$	2.71	s	-
$CH_3 - N_9$	3.46	s	-
H_A	4.28	dd	$J_{A,B} = 14.4$, $J_{A,H3} = 6.47$
H_B	4.63	dd	$J_{B,A} = 14.4$, $J_{B,H3} = 6.45$
H_6 and H_7	7.16-7.26	m	-
H_8	7.33	d	$J_{8,7} = 7.8$
$H_{4'}$*	7.44	d	$J_{4',5'} = 2.1$
$H_{5'}$*	7.60	d	$J_{5',4'} = 2.1$
H_5	7.97	d	$J_{5,6} = 7.2$

Figure 12. One-dimensional ^{13}C-NMR spectrum of ondansetron
hydrochloride.

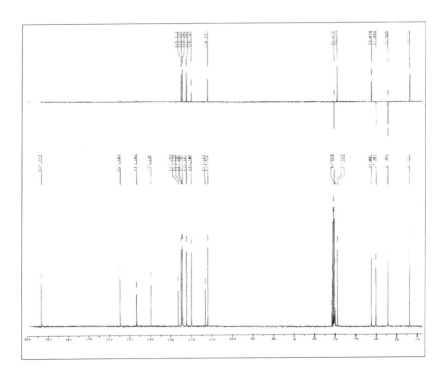

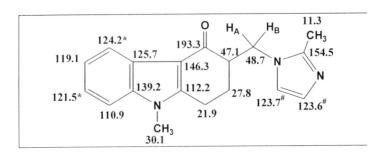

Figure 13.   Fast atom bombardment (FAB) mass spectrum of
ondansetron hydrochloride.

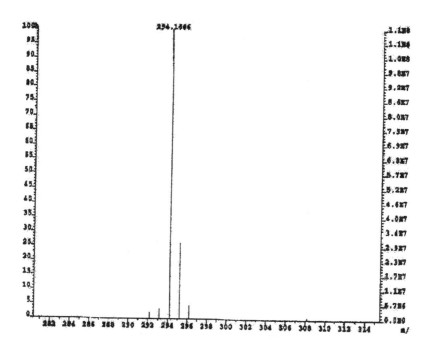

Figure 14. Electron impact (EI) mass spectrum of ondansetron
            hydrochloride.

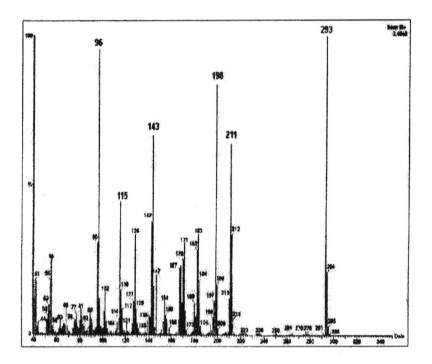

4. Methods of Analysis

4.1 Identification

The identity of ondansetron hydrochloride is most conveniently and quickly established using infrared absorption spectroscopy. The infrared spectrum is obtained by first mixing and grinding 1.5 mg of active substance with 100 mg of potassium bromide until a homogeneous blend is obtained, and then compressing a pellet from this mixture. The KBr pellet is then scanned between 400 to 4000 cm^{-1}. The identification may be based on the characteristic IR spectrum shown in Figure 9.

4.2 Titrimetric Analysis

Ondansetron hydrochloride can be assayed by non-aqueous titration. The method begins with the accurate weighing of 100 mg of the substance, diluting in 50 mL of glacial acetic acid, and finally adding 3 mL of acetic anhydride. The titrant is 0.1 N perchloric acid, and the endpoint is determined by potentiometric means. The percentage of analyte in the substance under test is obtained using:

$$\% \text{ OD} = [(36.59 \text{ V}) / \text{P}] [100]$$

where V is the volume (in mL) of 0.1 N perchloric acid consumed during the titration, and P is the sample weight taken.

4.3 Spectrophotometric Methods of Analysis

A spectrophotometric method for ondansetron determination in serum was reported [14]. The method is based on the formation of a 1:1 ion pair with bromcresol green in the pH range of 3.2 to 4.4. The quantitation was based on the UV absorbance at 420 nm. Linearity was demonstrated over the range of 0.1 - 20 mg/mL, with a relative standard deviation of 2.7%.

4.4 Chromatographic Methods of Analysis

4.4.1 Thin Layer Chromatography

A high performance TLC method was applied to the identification and quantitation of ondansetron hydrochloride. Samples having an analyte concentration of 1 mg/mL in water were prepared, and 10 μL spots were deposited onto HP-TLC silica gel plates. The plates were developed using 90:50:40:1 v/v chloroform-ethyl acetate-methanol-ammonia as the developing phase. After removal of the plates from the chamber and elimination of excess eluent, they were observed and scanned under 254 nm

UV light. Quantitation of the spots was carried out using a densitometric method. The sample spots coincided with those of the standard, having a Rf of about 0.73.

4.4.2   High Performance Liquid Chromatography

A sensitive HPLC method was developed to determine ondansetron hydrochloride in pharmaceuticals. A Waters Associates Inc.(Milford, MA, USA) liquid chromatograph was used to complete this study. The apparatus consisted of a model 600 quaternary pump and a model 996-photodiode-array detector. A pre-packed (5 μm, 25 cm x 3 mm I.D.) Hypersil CPS column (Nitrile groups chemically bonded to porous silica particles) was used, and a flow rate of 1.5 mL/min was found to give adequate resolution (back pressures on the column ranged between 2400 to 2500 p.s.i.). The temperature of the column was maintained at 40°C. Injection volumes of 5 μL were applied to the instrument by a model 717 Waters autosampler, and detection was made on the basis of the UV absorption at 215 nm. The mobile phase was prepared mixing 0.02 M monopotassium phosphate buffer with acetonitrile in the ratio of 80:20, with the pH then being adjusted to 5.4 with 0.1N NaOH. Samples and solvents were degassed and clarified by filtration through porous membranes (Millex-GV, 0.22 μm and Durapore 0.45 μm, both obtained from Waters). A typical chromatogram obtained using this method is shown in Figure 15.

Calibration curves were obtained by injecting serial dilutions of standard solutions of ondansetron hydrochloride in the concentration range of (0.10-0.40 mg/mL). Five determinations were obtained for each sample so that final values could be obtained from the mean of the five measurements. The analyses were run over two days, and linear line calibration plots were obtained, for which the correlation coefficient equaled or exceeded 0.9990 (r^2=99.8%). The variation coefficient of response factor (F) was 1.5%, and the relative standard deviation was = 1.3%. A study of the precision of the LC assay was performed over a 2-day period, calculating the standard deviation and variation coefficient for 5 determinations per day of two different solutions of ondansetron hydrochloride standards (0.25 mg/mL). The resulting coefficient of variation equaled 0.20 %, and the resolution factor to the nearest peak was 2.1%. The limit of detection was 0.01 μg/mL, and the limit of quantitation was 0.03 μg/mL. The tailing factor (T) was found to be 1.5%, and the column efficacy (N) was calculated to be 6400 theoretical plates. A photodiode array detector study proved that all chromatographic peaks were spectrally pure.

Figure 15.   Typical HPLC chromatogram with UV detection of
              ondansetron hydrochloride.

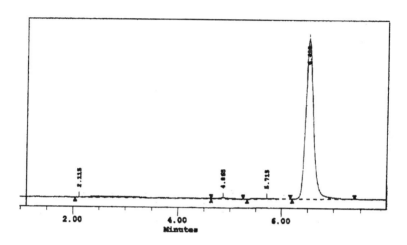

This method could also be applied to the determination of impurities and
degradation products, requiring slight modifications in the mobile phase
composition in some cases. For instance, occasionally the proportion of the
mobile phase should be switched to 90:10 buffer / organic modifier. This
method proved applicable to accelerated stability studies associated with
long-term stability studies.

A number of other HPLC methods have been reported to be suitable for the
identification and quantitation of ondansetron and its stability in
pharmaceuticals [15-17], as well, in biological fluids such as human plasma
or pharmacokinetics studies [18-19].

4.5   Determination in Body Fluids and Tissues

4.5.1   High Performance Liquid Chromatography

A stereospecific HPLC method was developed by Liu and Stewart [20] for
the assay of R-(–)- and S-(+)-ondansetron enantiomers in human serum.
The method involves the use of solid-phase extraction for sample clean-up,
and is also free of interference from 6-hydroxyondansetron, 7-
hydroxyondansetron, and 8-hydroxyondansetron (the three major

metabolites of ondansetron). Chromatographic resolution of the enantiomers was performed on a reversed-phase cellulose-based chiral column (Chiralcel OD-R) under isocratic conditions. The mobile phase consisted of 0.7 M sodium perchlorate-acetonitrile (65:35, v/v) eluted at a flow-rate of 0.5 mL/min, and detection was made on the basis of the UV absorbance at 210 nm. Linear calibration curves were obtained for each enantiomer in serum in the concentration range 15-750 ng/mL. The limit of quantitation of each enantiomer was 15 ng/mL, and the detection limit for each enantiomer in serum was 7 ng/mL. Recoveries were more than 90% for both ondansetron enantiomers.

Kelly et al. have developed a stereospecific, solid-phase extraction HPLC method with UV detection for the determination of the enantiomers of ondansetron in spiked human serum [21]. The method was suitable for the assay and quantitation of each enantiomer in the 10-200 ng/mL range.

4.5.2 Electrophoresis

As a useful alternative to existing chiral high-performance liquid chromatographic methods, a high-performance capillary electrophoresis assay method was reported for the quantitation of S-(+)- and R-(−)-ondansetron in human serum [22]. Resolution was achieved using 15 mM heptakis-(2, 6-di-O-methyl)-β-cyclodextrin in 100 mM phosphate buffer (pH 2.5). A 72-cm untreated fused-silica capillary, at a constant voltage of 20 kV, was used for the analysis. The analytes of interest were isolated from endogenous substances using a solid-phase extraction procedure. The detection limit was 10 ng/mL (using 2 mL of serum), and the limit of quantitation was 15 ng/mL. The calibration curve was linear over a range of 15-250 ng/mL, and the coefficients of determination obtained were greater than 0.999. Precision and accuracy of the method were 2.76-5.80 and 2.10-5.00%, respectively, for S-(+)-ondansetron, and 3.10-6.57 and 2.50-4.35%, respectively, for R-(−)-ondansetron.

4.5.3 Radioimmunoassay

Wring et al. have developed a sensitive radioimmunoassay, combined with solid-phase extraction, suitable for the subnanogram per milliliter determination of ondansetron in human plasma [23]. The antiserum was raised in Soay sheep following primary and booster immunizations with an immunogen prepared by conjugating 9-(carboxypropyl) ondansetron to bovine thyroglobulin. The radio-ligand consisted of ondansetron that had been specifically tritium-labeled on the N-methyl group of the indole moiety. The solid-phase extraction method, using a cyanopropyl sorbent, was introduced to remove cross-reacting metabolites and to enhance assay sensitivity. The calibration range was 0.05-2.40 ng/mL using a 1 mL sample of human plasma. Inter- and intra-assay bias and precision were less than ±13%, and less than 10% over this concentration range, respectively.

5.   Stability

The stability of ondansetron hydrochloride has been evaluated under different conditions, and excluding its chemical instability above pH =5.7, ondansetron hydrochloride has proved to be stable over the long term when stored in a tight-closed containers in the absence of light. The manufacturer states that ondansetron hydrochloride infusion solutions are stable at room temperature and when exposed to normal light for 48 hours [24].

In this context, and after evaluating the stability of 1.33 and 1 mg/mL injections of ondansetron hydrochloride in combination with 12 other injectable drugs, Stewart et al. reported that all solutions retained more than 90% of their initial ondansetron concentration [17]. The formulations were stored in plastic syringes or glass vials at 21.8-23.4°C and at 4°C for up to 24 hours.

Numerous studies have proved stability and compatibility of ondansetron hydrochloride with other drugs and in different kinds of containers [12, 16, 17, 25]. Leak and Woodford proved the stability of ondansetron for one month under daylight, although ondansetron hydrochloride was found to be unstable under intense light [10].

6.   Drug Metabolism and Pharmacokinetics

6.1   Bioavailability

Following oral administration, ondansetron hydrochloride is well absorbed and undergoes first-pass metabolism. The absolute oral bioavailability averages 67%. A single 8 mg dose, administered in either tablet or solution form, produces a peak plasma of about 0.03 to 0.04 µg/mL after 1.5 to 2 hours of administration [26].

The extent and rate of ondansetron absorption is greater in women than men, and food may increase the bioavailability of ondansetron. The absorption from tablet dosage forms is not dissolution rate limited, with

Bozigian *et al.* having found that 100% of ondansetron tablets dissolved *in vitro* within 30 minutes, which was consistent with their *in vivo* findings. These workers also proved the bioequivalency between 8 mg tablets and an oral solution [27].

Ondansetron is also well absorbed when administrated rectally [28]. Suppository formulations of 16 mg closely approximated the pharmacokinetics of the oral tablets, with plasma AUC values being comparable to those obtained after an 8 mg oral administration.

Hsyu *et al.* have evaluated ondansetron absorption after intravenous infusion, oral dose, infused to the colon via nasogastric intubation, and administered to the rectum by retention enema [29]. Peak maxima and T_{max} values were not different for the various routes of administration. They concluded that ondansetron is well absorbed in all studied intestinal segments.

Chemotherapy does not alter the pharmacokinetics of oral ondansetron in cancer patients during the treatment, and the mean bioavailability of this drug appears to be greater in cancer patients than in healthy volunteers due to a reduced first-pass metabolism of ondansetron [30].

The pharmacokinetics of ondansetron was also investigated in patients with varying degrees of hepatic impairment, and compared to young healthy volunteers [31]. After a single 8 mg intravenous dose was given over 5 minutes, patients with severe hepatic impairment had a lower mean plasma clearance (96 mL/min *vs.* 478 mL/min in young healthy volunteers). In addition, increased AUC values (1383 ng/mL hour vs. 279 ng/mL hour) and T_{max} (21 hour *vs.* 3.6 hour) were observed. It was concluded that a reduction in dosing frequency is advisable in patients with severe hepatic impairment, and a once-daily dose was suggested as the most appropriate in these cases.

6.2   Distribution

Ondansetron has a large volume of distribution of about 163 L [32]. The volume of distribution in healthy young males following administration of 8 mg of ondansetron as an intravenous infusion over 5 minutes was about 160 L. Patients of 4-12 years in age reportedly have a volume of distribution somewhat larger relative to adults. Practically 36% of circulating ondansetron is distributed into erythrocytes, and protein binding is reported to be moderate (about 70-76%) [31].

6.3   Metabolism

Ondansetron is subject to extensive metabolism, primarily hydroxylation on the indole ring that is followed by glucoronide or sulfate conjugation. *N*-demethylation is a minor route of metabolism. 8-hydroxy-ondansetron is an active metabolite of ondansetron, and has an antiemetic activity equivalent to the parent compound. Approximately 5% of the dose is excreted as unchanged parent compound in the urine [1], and metabolites are excreted in urine and feces.

After studying the pharmacokinetics of ondansetron hydrochloride dihydrate administrated intravenously to healthy subjects (who had been previously phenotyped with debrisoquin (debrisoquine) as 6 poor metabolizers and 6 extensive metabolizers], Ashforth *et al.* concluded that ondansetron clearance is not mediated exclusively by cytochrome P-450 2D6 [33]. They learned that there was no significant difference in AUC, maximum concentration, clearance, or half-life between poor and extensive metabolizers.

6.4   Elimination

Following 8 mg I.V. administration of ondansetron, Rojanasthien *et al.* reported that the mean value of ondansetron elimination half-life is about 4.5 hours, and the plasma clearance is 398 mL/min [34]. Elimination is predominantly renal, and less than 5% of an intravenous dose of ondansetron is recovered unchanged in the urine.

The mean elimination half-life in healthy young volunteers is approximately 3 hours, but is 4 hours in cancer patients. Elderly patients tend to have an increased elimination half-life, while most pediatric patients less than 15 years of age have shorter plasma half lives (about 2.4 hours) than do patients older than 15 years of age [35]. In patients with severe hepatic impairment, clearance is decreased and the apparent volume of distribution of ondansetron is increased with a resultant increase in plasma half-life [36]. The manufacturer recommends that the total daily dose not exceed 8 mg in such patients [24].

6.5   Dose Schedule

The usual I.V. adult and adolescent dose is 0.15 mg per kg of body weight [1], and administration is repeated at four and eight hours following the first dose. Used as prophylaxis for nausea and vomiting associated with cancer chemotherapy, the drug may be administered orally (8 mg) 30 minutes prior to treatment. In post chemotherapy, 8 mg dosing is recommendable. In children younger than four years, a dosage has not been established. In instances of hepatic function impairment, the maximum recommended dose of ondansetron is 8 mg per day.

7.   Pharmacology

7.1   Mechanism of Action

Chemotherapy in general, and specially cisplatin, produces nausea and vomiting via increasing the release and excretion of 5-hydroxy-indole acetic acid (5-HIAA), a stable metabolite of serotonin. Vomiting and nausea coincide with peak secretion of 5-HIAA [37]. Very high densities of 5-HT3 receptors are located on vagal nerve terminals and in part of the central nervous system called the area postrema, where the chemoreceptor trigger zone is located. It is not clear whether the antiemetic action of ondansetron is mediated centrally, peripherally, or in both sites, but it is suggested that chemotherapy causes the release of serotonin in the gastric tract that stimulates vagal afferent fibers and initiates sensory signals to the chemoreceptor trigger zone inducing nausea and vomiting [37,38]. Ondansetron is a potent, highly selective, competitive antagonist at the 5-HT3 receptor, and thus inhibits the symptoms of nausea and vomiting.

On the other hand, Martin et al. proposed that radiation induces emesis by a dual mechanism consisting of an early peripheral mechanism and a later central mechanism [39]. They found out that two deliveries of 5-HT3 receptor antagonists seem to disrupt serotonergic transmission at the brain stem structures, and affects the peripheral release of serotonin from the gut, completely preventing radiation-induced vomiting. Their study confirms that the 5-HT3 dependent mechanisms that mediate emesis are similar for both neutron and gamma radiation.

Compared with metoclopramide, ondansetron hydrochloride demonstrates equal or superior efficacy, but has no dopamine-receptor antagonist activity, and thus does not induce extrapyramidal side effects [26].

A similar efficacy of prophylactically administered ondansetron was found relative to droperidol for the reduction of postoperative nausea associated with laparoscopic surgery in female patients [40]. However, ondansetron appears to be slightly more efficient than droperidol in preventing vomiting.

7.2 Toxicity

Ondansetron is well tolerated, and has fewer side effects than does metoclopramide, owing probably to a lack of activity by ondansetron at dopamine receptors.

Preclinical studies demonstrated that there is no end-organ toxicity in rats and dogs administered ondansetron doses 30 to 100 times those used in humans [41]. At near-lethal doses of ondansetron, animals developed subdued activity, ataxia, and convulsions.

Earlier studies performed by Cunningham *et al.* showed that the only adverse effects were dryness of the mouth, mild sedation, and diarrhea, and that these were not clearly related to the drug [42]. Other studies indicated that incidence of headache may be dose related [41].

Diarrhea is the most common adverse GI effect of ondansetron [41,43], occurring in 8-16% of patients receiving the I.V. drug in recommended dosages concomitant with cisplatin. However, since cisplatin can cause diarrhea, a causal relationship to ondansetron has not been established.

Constipation occurs in 5-7% of patients receiving multiple-day oral therapy in the recommended dosage for chemotherapy-induced nausea and vomiting [36]. The incidence of constipation may be dose related [44]. Other adverse GI effects include abdominal pain in 4-5% of patients [43], and xerostomia, dyspepsia or heartburn, abnormal taste and anorexia [45].

Because most patients receiving ondansetron in clinical trials for chemotherapy-induced nausea and vomiting had serious underlying disease (*e.g.*, cancer) and were receiving toxic drugs (*e.g.*, cisplatin), diuretics, and I.V. fluids concomitantly, it may be difficult to attribute various adverse effects to ondansetron [24, 36, 46, 47].

Insufficient information is available regarding dosage in children younger than 3 years, and more data about safety and efficacy of ondansetron in children is to be collected [24, 36, 48].

No evidence has been reported regarding carcinogenicity or mutagenicity [24, 36, 41], but the drug should be used during pregnancy only when clearly needed [24, 36].

References

1.   *USP DI*, 15th edn. The United States Pharmacopoeial Convention
      Inc., Rockville, MD, 1995, pp. 2062-2063.

2.   A. Pinelli, S. Trivulzio, and L. Tomasoni, *Eur. J. Pharmacol.*, **340**,
      111-119 (1997).

3.   I.H. Coates, J.A. Bell, and D.C. Humber, GB Appl. 85/1,727
      (1985); Eur. Pat. Appl. EP 191,562 (1986); *Chem. Abstr.*, **105**:
      226579u (1986).

4.   I.H. Coates, J.A. Bell, and D.C. Humber, GB Appl. 85/18,743
      (1985); Eur. Pat. Appl. EP 219,193 (1987); *Chem. Abstr.*, **107**:
      176032d (1987).

5.   N. Godfrey, I.HJ. Coates, and J.A. Bell, Eur. Pat. Appl. EP 221,629
      (1987); *Chem. Abstr.*, **107**: 77803z (1987).

6.   J.A. Bell, I.H. Coates, and C.D. Eldred, Eur. Pat. Appl. EP 219,929
      (1987); *Chem. Abstr.*, **107**: 77804a (1987).

7.   M.Y. Kim, G.J. Lim, and J.I. Lim, *Heterocycles*, **45**, 2041-2043
      (1997).

8.   K.C. Mohan and K. Ravikumar, *Acta Cryst., Sect. C, Cryst. Struct.
      Comm.*, **C51**, 2627 (1995).

9.   *Merck Index*; 20th Edition, Merck & Co., Rahway, NJ, 1996, p.
      6799.

10.   R.E. Leak and J.D. Woodford, *Eur. J. Cancer Clin. Oncol.*, **25**
      (Suppl 1), S67 (1989).

11.   J.W. MacKinnon and D.T. Collin, *Eur. J. Cancer Clin. Oncol.*, **25**
      (Suppl 1), S61 (1989).

12.   R.A. Fleming, D.J. Olsen, and P.D. Savage, *Am. J. Health Syst.
      Pharm.*, **52**, 514 (1995).

13.   P.F. Jarosinski and S. Hirschfield, *New. Engl. J. Med.*, **325**, 1315
      (1991).

14.   L. Zamora and J. Calatayud, *Anal. Lett.*, **29**, 785 (1996).

15.   L. Ye and J.T. Stewart, *J. Liquid Chrom. Rel. Tech.*, **19**, 711 (1996).

16. B. Evrard, A. Ceccato, and O. Gaspard, *Am. J. Health Syst. Pharm.*, **54**, 1065 (1997).

17. J.T. Stewart, F.W. Warren, and D.T. King, *Am. J. Health Syst. Pharm.*, **55**, 2630 (1998).

18. P.V. Colthup, C.C. Felgate, and J.L. Palmer, *J. Pharm. Sci.*, **80**, 868 (1991).

19. M. Depot, S. Leroux, and G. Caille, *J. Chrom. Biomed. Appl.*, **693**, 399 (1997).

20. J. Liu and J.T. Stewart, *J. Chrom.. Biomed. Appl.*, **694**, 179 (1997).

21. J.W. Kelly, L. He, and J.T. Stewart, *J. Chrom. Biomed. Appl.*, **133**, 291 (1993).

22. M. Siluveru and J.T. Stewart, J. Chrom. Biomed. Appl., 691, 217 (1997).

23. S.A. Wring R.M. Rooney, C.P. Goddard, I, Waterhouse, and W.N. Jenner, *J. Pharm. Biomed. Anal.*, 12, 361 (1994).

24. "Zofran (Ondansetron Hydrochloride) Injection Prescribing Information", Glaxo-Wellcome, Research Triangle Park, NC, 1994.

25. R.L. Hagan, *Am. J. Health Syst. Pharm.*, **54**, 1110 (1997).

26. ***Martindale, The Extra Pharmacopoeia***, 31st edn., E.F. Reynolds, ed., Royal Pharmaceutical Society. London, 1996, p. 1233.

27. H.P. Bozigian, J.F. Pritchard, A.E. Gooding, *J. Pharm. Sci.*, **83**, 1011 (1994).

28. M.W. Jann, T.L. ZumBrunnen, S.N. Tenjarla, *Pharmacotherapy*, **18**, 288 (1998).

29. P.H. Hsyu, J.F. Pritchard, H.P. Bozigian, *Pharm. Res.*, **11**, 156 (1994).

30. P.H. Hsyu, J.F. Pritchard, H.P. Bozigian, *J. Clin. Pharmacol.*, **34**, 767 (1994).

31. J.C. Blake, J.L. Palemer, and N.A. Minton, *J. Clin. Pharmacol.*, **35**, 441 (1993).

32.   P.V. Colthup and J.L. Palmer, *Eur. J. Cancer Clin. Oncol.*, **25** (Suppl 1), S71 (1989).

33.   E.I. Ashforth, J.L. Palmer, and A. Bye, *Br. J. Clin. Pharmacol.*, **37**, 389 (1994).

34.   N. Rojanasthien, M. Manorot, and B. Kumsorn, *Int. J. Clin. Pharmacol. Therap.*, **37**, 548 (1999).

35.   J.F. Pritchard, J.C. Bryson, and A.E. Kernodle, *Clin. Pharmacol. Therap.*, **41**, 51 (1992).

36.   ***Physicians' Desk Reference***, 50th edn., Medical Economics Co. Inc., Montvale, 1996, pp. 12, 17-19.

37.   L.X. Cubeddu, I.S. Hoffmann, N.T. Fuenmayor, *New Engl. J. Med.*, **322**, 810 (1990).

38.   M.B. Tyres, K.T. Bunce, and P.P.A. Humphrey, *Eur. J. Cancer Clin. Oncol.*, **25** (Suppl 1), S15 (1998).

39.   C. Martin, V. Roman, and D. Agay, *Radiation Res.*, **149**, 631 (1998).

40.   M. Koivuranta, E. Laara, P. Ranta, *Acta Anaesthesiol. Scand.*, **41**, 1273 (1997).

41.   A.L. Finn, *Semin. Oncol.*, **19** (Suppl. 10), 53 (1992).

42.   D. Cunningham, J. Hawthorn, A. Pople, *Lancet*, **1**, 1461 (1987).

43.   J.C. Bryson, *Semin. Oncol.*, **19** (Suppl. 15), 26 (1992).

44.   L.X. Cubeddu, K. Pendergrass, T. Ryan, *Am. J. Clin. Oncol.*, **17**, 137 (1994).

45.   P.J. Hesketh, W.K. Murphy, E.P. Lester, *J. Clin. Oncol.*, **7**, 700 (1989).

46.   R.J. Milne and R.C. Heel, *Drugs*, **41**, 574 (1991).

47.   W.M. Castle, A.J. Jukes, C.J. Griffiths, *Eur. J. Anaesthesiol.*, **9** (Suppl. 6), 63 (1992).

48.   K.D. McQueen and J.D. Milton, *Ann. Pharmacotherap.*, **28**, 85 (1994).

SILDENAFIL CITRATE

Adnan A. Badwan, Lina Nabulsi, Mahmoud M. Al-Omari,
Nidal Daraghmeh, and Mahmoud Ashour

The Jordanian Pharmaceutical Manufacturing Co.
Naor P.O. Box 94
Amman, Jordan

339

Contents

1. Description

1.1 Nomenclature

1.1.1 Chemical Name

Piperazine,1-[[3-(6,7-dihydro-1-methyl-7-oxo-3-propyl-1*H*-pyrazolo[4,3-*d*]-pyrimidin-5-y1)-4-ethoxyphenyl]sulfonyl]-4-methyl-,2-hydroxy-1,2,3-propanetricarboxylate (1:1).

1-[[3-(6,7-Dihydro-1-methyl-7-oxo-3-propyl-1*H*-pyrazolo[4,3-*d*]pyrimidin-5-y1)-4-ethoxyphenyl]sulfonyl]-4-methylpiperazine citrate [8].

1.1.2 Nonproprietary Names

Sildenafil citrate; UK 92480

1.1.3 Proprietary Names

Sildenafil citrate is marketed by Pfizer under the proprietary name, Viagra®.

1.2 Formulae

1.2.1 Empirical

$C_{22}H_{30}N_6O_4S \cdot C_6H_8O_7$

1.2.2 Structural

1.3   Molecular Weight

666.71

1.4   CAS Number

171599-83-0; 139755-83-2

1.5   Appearance

Sildenafil citrate is obtained as a white to off-white crystalline and odorless powder.

1.6   Uses and Applications

Sildenafil citrate is a compound of the pyrazolo-pyrimidinyl-methylpiperazine class, and is used to treat male erectile dysfunction. It is a selective inhibitor of cyclic guanosine monophosphate (cGMP)-specific phophodiesterase type 5 (PDE5) [1]. Sildenafil citrate was discovered by scientists at Pfizer USA, and was approved by the Food and Drug Administration on March 27, 1998 [1].

2.   Method of Preparation

The preparation of Sildenafil citrate was reported by Terrett *et al.* in 1996 [2]. The reactions involved in this synthetic pathway are shown in Scheme 1, and the synthesis was carried out according to the following route. A pyrazole ring was formed by reacting diketoester **(1)** and hydrazine. This was followed by the regioselective N-methylation of the pyrazole **(2)**, and hydrolysis gave the carboxylic acid **(3)**. The resulting compound was nitrated **(4)**, and this was followed by carboxamide formation and nitro group reduction to give the pyrazole intermediate **(5)**. The amine group was acylated with 2-substituted benzoyl chloride **(6)**, and then by cyclization under basic conditions to produce the pyrazolopyrimidinone **(7)**. Subsequently, chlorosulfonylation **(8)**, followed by coupling with piperazine, yielded Sildenafil. The latter was reacted with citric acid to produce Sildenafil citrate.

Scheme 1. Synthetic route for the preparation of Sildenafil citrate.

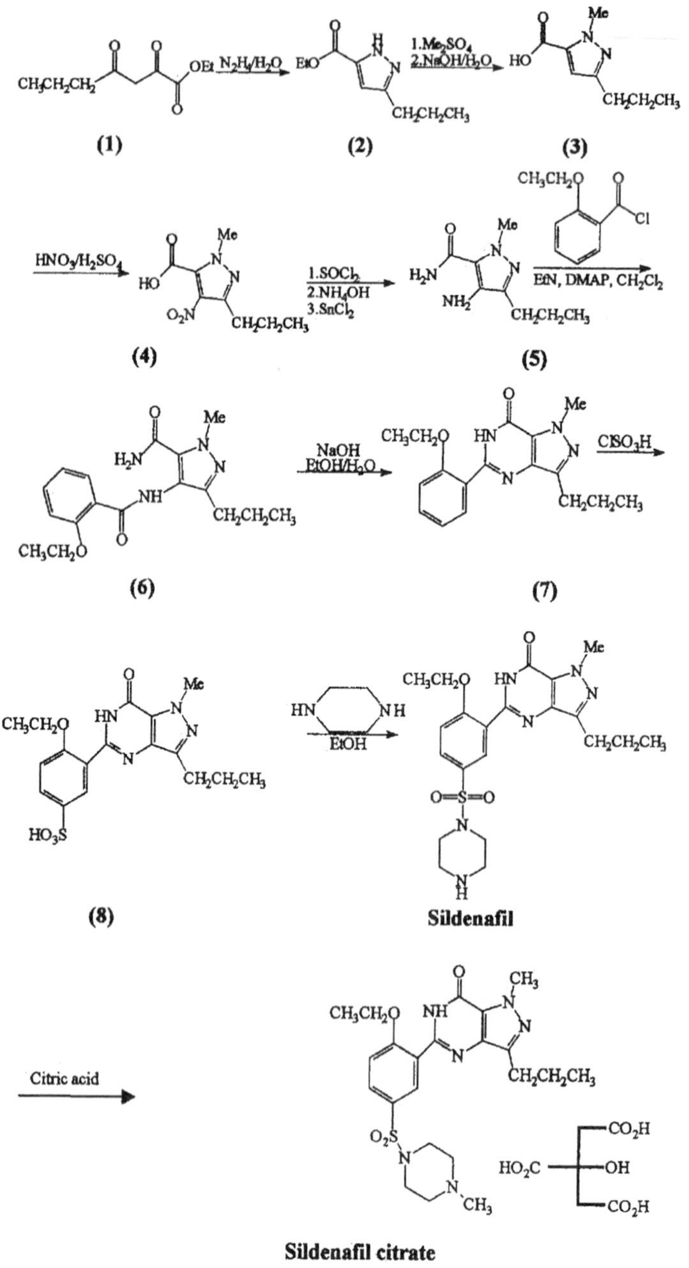

(1)

(2)

(3)

(4)

(5)

(6)

(7)

(8)

Sildenafil

Sildenafil citrate

3. Physical Properties

3.1 X-Ray Powder Diffraction Pattern

The X-ray powder diffraction pattern of Sildenafil citrate was obtained using a Philips diffractometer system (model PW 105-81 Goniometer and PW 1729 Generation) [3]. The pattern was obtained using nickel filtered copper radiation (λ = 1.5405 Å), and is shown in Figure 1. A full data summary is compiled in Table 1.

3.2 Thermal Methods of analysis

3.2.1 Differential Scanning Calorimetry

The thermal behavior of Sildenafil citrate was examined by DSC, using a TA Instruments 910S differential scanning calorimeter. The system was calibrated with Indium. Sildenafil citrate samples ranging in size from 5 to 10 mg were run at a scanning rate of 5°C/min. over a temperature range of 30° to 350°C. Thermograms of Sildenafil citrate showed one endothermic feature, attributable to fusion, exhibiting a peak maximum at 199.4°C. As evident in the reference DSC of Figure 2, the melting process appeared to take place with decomposition [3].

3.2.2 Thermogravimetric Analysis

Thermogravimetric thermograms were obtained using model 951 TA Instruments system. The system was calibrated using Indium, and the experiments were carried out at heating rate of 10°C/min. The sample size ranged between 5 to 10 mg, and samples were analyzed over a temperature range of 30°C to 350°C. The TG thermogram of reference Sildenafil citrate is shown in Figure 3, illustrating the anhydrous nature of the compound in that no weight loss is noted below a temperature of 200°C. The weight loss of 26.5% takes place at a temperature above the melting point of Sildenafil citrate (195°C), and the additional decomposition is associated with pyrolysis of the compound [3].

3.2.3 Thermal Evidence for Polymorphism

Recrystallization of Sildenafil citrate from different solvents led to non-equivalent DSC thermograms that were interpreted to demonstrate the existence of three polymorphs. As shown in Figure 3, one may observe thermally induced conversion of one form into another during the heating

Figure 1.   X-ray powder diffraction pattern of Sildenafil citrate.

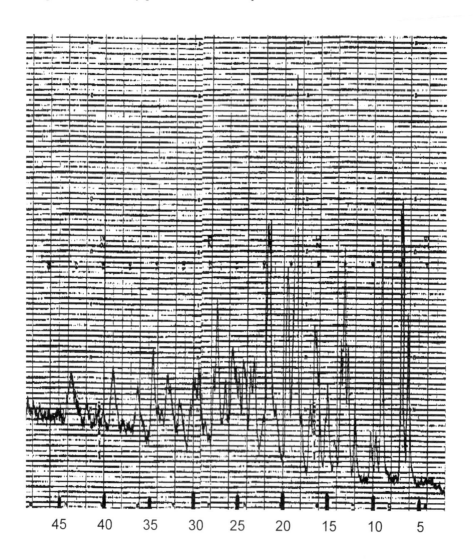

Scattering Angle (degrees 2-θ)

Table 1

Crystallographic Data Deduced from the X-Ray Powder Pattern of
Sildenafil Citrate

Scattering Angle (degrees 2-θ)	d-spacing (Å)	Relative Intensity (%)	Scattering Angle (degrees 2-θ)	d-spacing (Å)	Relative Intensity (%)
6.2	14.255	42.1	23.2	3.8338	25.3
6.8	12.998	70.5	23.6	3.7697	21.0
9.1	9.7176	62.1	24.4	3.6479	24.2
9.7	9.1178	11.6	24.8	3.5900	20.0
10.3	8.5880	11.6	25.2	3.5339	22.1
12.0	7.3750	17.9	25.7	3.4662	26.3
12.8	6.9157	32.6	26.4	3.3759	16.8
13.2	6.7071	56.8	27.4	3.2549	26.8
13.6	6.5107	33.7	27.7	3.2208	18.9
15.2	5.8288	21.0	29.5	3.0228	21.0
16.1	5.5049	23.2	30.0	2.9785	17.9
16.3	5.4378	34.7	31.6	2.8312	9.5
16.6	5.3402	35.8	32.8	2.7303	15.8
17.0	5.2154	13.7	34.5	2.5256	22.1
18.6	4.7702	100.0	36.2	2.4813	13.7
19.6	4.5291	48.4	39.0	2.3093	15.8
21.4	4.1520	60.0	43.7	2.0713	12.6
21.8	4.0767	58.9			

A.A. BADWAN, L. NABUSLI, M.M. AL-OMARI,
N. DARAGHMEH, AND M. ASHOUR

Figure 2.   Differential scanning calorimetry thermogram of Sildenafil
citrate.

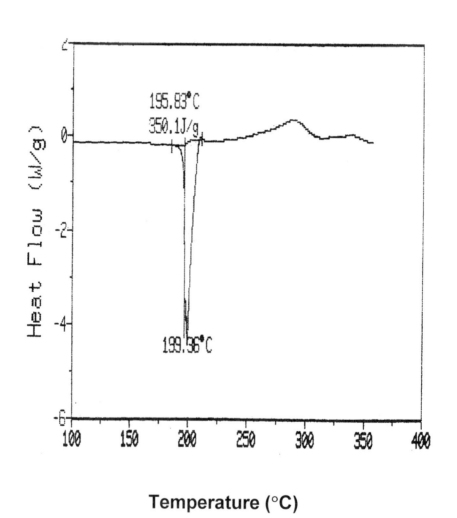

Temperature (°C)

Figure 3.   Thermogravimetric analysis of Sildenafil citrate.

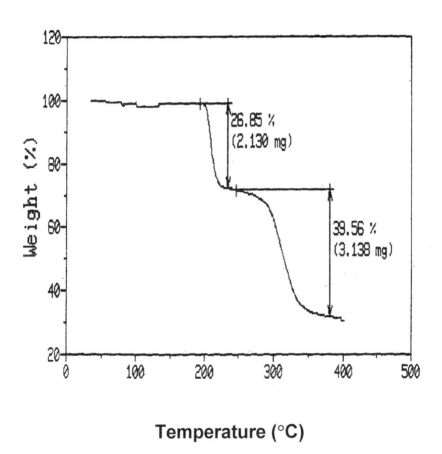

Figure 4.   Differential scanning calorimetry thermogram of a sample of
Sildenafil citrate illustrating the conversion from
polymorph II to polymorph I.

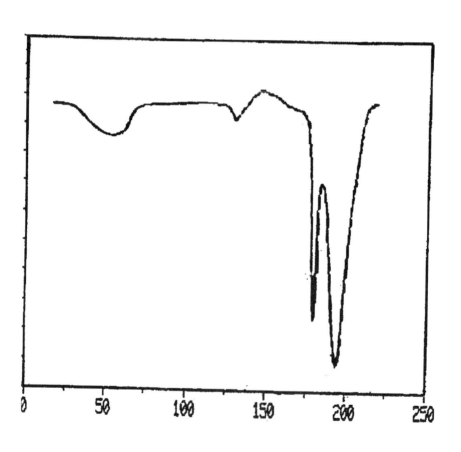

Temperature (°C)

Table 2

Characterization of the Different Polymorphs of Sildenafil Citrate

Solvent System	Crystallization Conditions (from hot solution)	Melting point (DSC Onset, °C)	Heat of fusion (ΔH, J/g)	Form Obtained
Water	Slow cooling	199.0	333.1	I
Water: methanol (1:1)	Slow cooling	199.3	344.5	I
Water: ethanol (1:1)	Slow cooling	199.3	335.0	I
Water: methanol (2:1)	Slow cooling	199.5	327.9	I
Water: ethanol (2:1)	Slow cooling	199.4	339.7	I
Acetone	Slow cooling	195.3	348.5	II
Ethanol	Slow cooling	193.7	326.5	II
Methanol	Slow cooling	195.3	339.3	II
Propylene glycol	Rapid cooling	192.5	305.4	II
Isobutanol	Slow cooling	194.7	384.7	II
1-propanol	Slow cooling	195.3	372.4	II
2-propanol	Rapid cooling	194.8	365.7	II
1,4-dioxane	Slow cooling	194.2	344.5	II
Ethanol	Rapid cooling	186.0	309.2	III

process, with the endotherms being easily distinguishable. A summary of the various polymorphic forms of Sildenafil citrate obtained as a function of solvent and crystallization conditions is shown in Table 2 [4]. It was found that polymorph I melts in the range of 199.0-199.5°C, polymorph II melts in the range of 192.5-195.3°C, and polymorph III melts at 186.0°C.

3.3 Ionization Constants

Sildenafil has a basic functional group, characterized by a pK value of 8.7 (NH-piperazine). In addition, it also has a weak acidic moiety (HN-amide) [4].

Aqueous solutions of Sildenafil citrate are acidic. The pH of 0.3 g/100 mL aqueous solution is 3.7, as measured by a glass electrode [3].

3.4 Solubility Characteristics

The solubility of Sildenafil citrate has been determined in a variety of different solvents, and the results of this study [3] are shown in Table 3.

The pH solubility profile of Sildenafil base was established in 0.05 M citrate buffer. Excess amounts of the drug were added to 50 mL of the desired medium. The samples were shaken in a thermostat shaker at about 200 rpm at 30°C for about 24 hours to attain equilibrium, and then an aliquot was centrifuged (if necessary) and filtered using a 0.45 μm filter. The content of the solutions was assayed spectrophotometrically at 290 nm using a Beckman (USA) DU-650i spectrophotometer, and these results are collected in Table 4. These indicated that solubility decreases with increasing pH, followed by an increase above pH 9. Figure 5 graphically depicts the pH solubility profile of Sildenafil base [3].

3.5 Partition Coefficients

The partition coefficient of Sildenafil citrate between *n*-octanol and water was spectrophotometrically determined at room temperature, reading the solutions at 290 nm, using a Beckman DU 650i spectrophotometer. The partition coefficient was found to be 0.78 [3].

Table 3

Solubility of Sildenafil Citrate in Different Solvents at 25°C

Solvent	Quantity Dissolved (mM)
Methanol	1.20
Water	0.60
Ethanol	0.17
Dichloromethane	0.005
Diethyl ether	0.005
n-Hexane	0.0001

Figure 5.   pH solubility profile of Sildenafil base.

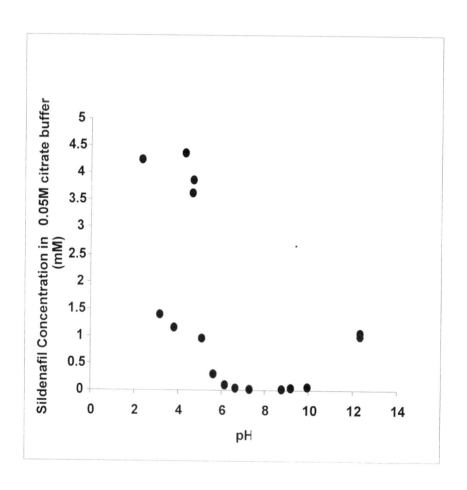

Table 4

pH Solubility Profile of Sildenafil Citrate at 30°C

pH	Quantity Dissolved (mM)
2.35	4.25
3.18	1.40
3.81	1.16
4.33	4.35
4.66	3.63
5.1	0.96
5.65	0.30
6.15	0.10
7.29	0.027
8.77	0.026
9.2	0.032
9.97	0.055
12.32	1.05

3.6   Spectroscopy

3.6.1   UV/VIS Spectroscopy

The UV spectrum of Sildenafil citrate was obtained using a Beckman DU-650i spectrophotometer system. To obtain the spectrum shown in Figure 6, the substance was dissolved in 30% methanolic water [3].

3.6.2   Vibrational Spectroscopy

The infrared spectrum of Sildenafil citrate was obtained as a KBr pellet using a Nicolet Impact 410 Infrared spectrophotometer, and is shown in Figure 7. Assignments for the observed spectral peaks to various molecular vibrations have been deduced, and these are presented in Table 5 [3].

3.6.3   Nuclear Magnetic Resonance Spectrometry

3.6.3.1   ^{1}H-NMR Spectrum

The one-dimensional proton ^{1}H-NMR spectrum of Sildenafil base dissolved in $CDCl_3$ is shown in Figure 8, and Table 6 lists the corresponding spectral assignments [3].

3.6.3.2   ^{13}C-NMR Spectrum

Figure 9 shows the one-dimensional ^{13}C-NMR spectrum of Sildenafil dissolved in $CDCl_3$, which was recorded at 24°C and internally referenced to TMS. The ^{13}C-NMR spectral assignments are presented in Table 7 [3].

3.6.4   Mass Spectrometry

The characteristic mass spectrum of Sildenafil base is shown in Figure 10, and the mass fragments with their assigned structures are shown in Table 8 [3].

Figure 6.   Ultraviolet absorption spectrum of a 3 mg/100 mL solution
             Sildenafil citrate in 30:70 v/v methanol / water.

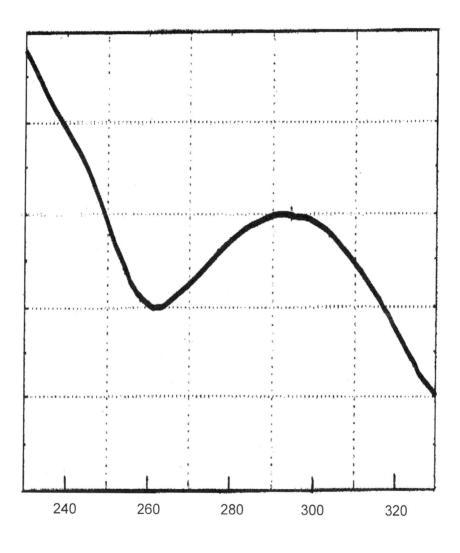

Wavelength (nm)

Figure 7.   Infrared absorption spectrum of Sildenafil citrate
              (KBr pellet).

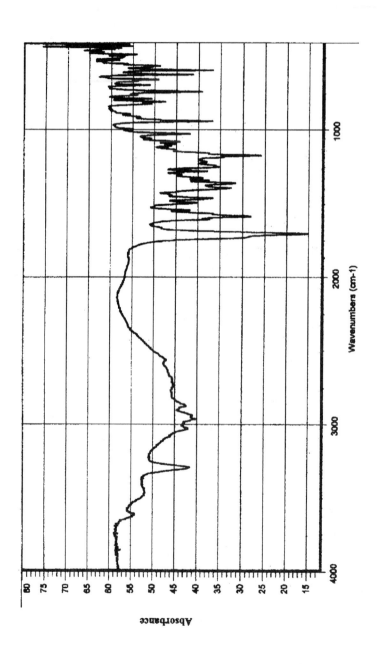

Table 5

Assignments for the Infrared Absorption Bands of Sildenafil
Citrate

Energy (cm^{-1})	Assignment
3617	O-H (Stretching)
3300	N-H (Stretching)
3025	C-H (Stretching Aromatic)
3000-2270	C-H (Stretching aliphatic)
1700	C=O (Stretching)
1600-1500	C=C (Aromatic)
1358,1173	SO_2 (Stretching)
1252	C-N (Stretching)

Figure 8. ^{1}H-NMR spectrum and numbering system of Sildenafil base in CDCl$_3$.

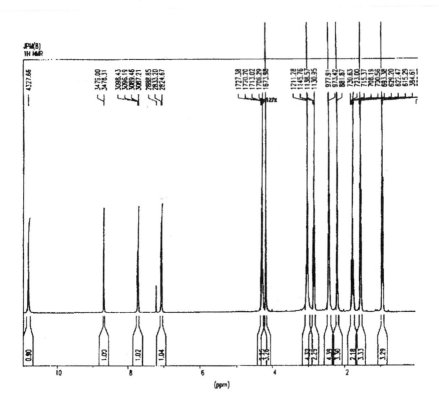

Table 6

Assignments for the ^{1}H-NMR Spectrum of Sildenafil Base

Chemical Shift (ppm)	Multiplicity	Proton Assignment
0.90	3(t)	A
1.56	3(t)	B
1.78	2(m)	C
2.24	3(s)	D
2.43	4(t)	E
2.85	2(t)	F
3.03	4(t)	G
4.15	3(s)	H
4.29	2(q)	I
7.07	1(d)	J
7.73	1(d)	K
8.69	1(d)	L
10.83	1(s)	M

Figure 9. ^{13}C-NMR spectrum and numbering system of Sildenafil base
in CDCl$_3$.

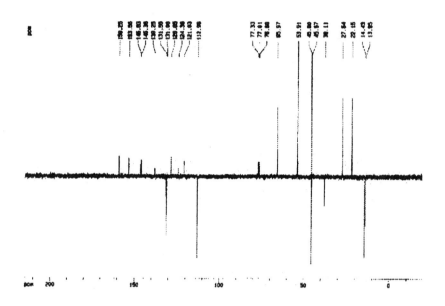

Table 7

Assignments for the ^{13}C-NMR Spectrum of Sildenafil Base

Carbon number	Chemical Shift (ppm)		Carbon number	Chemical Shift (ppm)
A	13.95		k	121.01
b	14.43		l	124.39
c	22.15		m	128.65
d	27.64		n	131.00
e	45.57		o	131.55
f	45.80		p	138.25
G	53.91		q	146.36
h	65.97		r	146.83
I	76.69		s	153.56
J	112.90		t	159.25

A.A. BADWAN, L. NABUSLI, M.M. AL-OMARI,
N. DARAGHMEH, AND M. ASHOUR

Figure 10.   Mass spectrum of Sildenafil citrate.

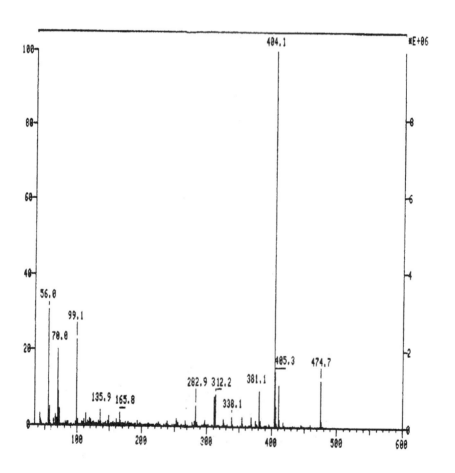

Table 8

Mass Spectrum Assignments for Sildenafil Citrate

m/z	Assignment	Structure
474.7	$C_{22}H_{30}N_6O_4S^+$	
404.1	$C_{18}H_{23}N_5O_4S^+$	
312.2	$C_{17}H_{20}N_4O_2^+$	
282.9	$C_{13}H_{19}N_2O_3S^+$	
99.1	$C_5H_{11}N_2^+$	
70.0	$C_4H_8N^+$	
56.0	$C_3H_6N^+$	

4.   Methods of Analysis

4.1   Identification

4.1.1   Infrared spectrum

Sildenafil citrate may be identified through an equivalence of the
absorption spectrum of the analyte with the characteristic infrared
absorption spectrum (KBr pellet method) described in section 3.6.2.

4.1.2   Thin layer chromatography

In the thin layer chromatography test described in section 4.4.1, the
principal spot in the chromatogram obtained is similar in position, color,
and size to that in the reference standard chromatogram.

4.2   Titrimetric Analysis

4.2.1   Non-aqueous Titration

Sildenafil citrate (250 mg) is dissolved in 50 mL of glacial acetic acid,
titrated with 0.1 M perchloric acid, with the end point being detected
potentiometrically. Each milliliter of 0.1 M perchloric acid VS is
equivalent to 66.67 mg of Sildenafil citrate, calculated as $C_{22}H_{30}N_6O_4S \cdot$
$C_6H_8O_7$. This method is suitable to assay Sildenafil citrate in its bulk drug
form, and in formulated products [3].

4.2.2   Aqueous Titration

In this method, 160 mg of Sildenafil citrate was accurately weighed and
dissolved in 50 mL of a 20:80 mixture of methanol / water. Titration
against 0.1 N sodium hydroxide was carried out using a Mettler DL-67
potentiometric titrator. Each milliliter of 0.1 N sodium hydroxide VS is
equivalent to 22.22 mg of Sildenafil citrate, calculated as $C_{22}H_{30}N_6O_4S \cdot$
$C_6H_8O_7$ [3].

4.3   Spectrophotometric Methods of Analysis

Dissolve an amount of Sildenafil citrate equivalent to 50 mg Sildenafil in
30% v/v aqueous methanol, and dilute quantitatively to 100 mL using the
same solvent. Sonicate the mixture for 20 minutes, and then centrifuge at

4000 rpm for 10 minutes. A 2 mL aliquot of the supernatant layer is further diluted to 50 mL with the same solvent, and the solution is scanned to determine the maximum absorption in the vicinity of 290 nm. The molar absorptivity at this maxima is about 295 [3].

4.4 Chromatographic Methods of Analysis

4.4.1 Thin Layer Chromatography

A thin layer chromatographic method was developed using silica gel 60 as the stationary phase. The mobile phase consisted of a mixture prepared by combining 50 volumes of 95% ethanol 50 volumes of dichloromethane, and 5 volumes of triethylamine. Separately apply 50 μL of a solution containing 0.5% w/v of Sildenafil citrate in methanol, develop the plate and after its removal, allow it to dry in air and examine under ultraviolet light (254 nm). This method can be used for the determination of Sildenafil citrate, 1-methyl-4-nitro-3-*n*-propyl-5-pyrazole carboxamide, as well as other impurities. The Rf for Sildenafil citrate is 0.77, and the Rf for 1-methyl-4-nitro-3-*n*-propyl-5-pyrazole carboxamide is 1.1 [3].

4.4.3 High Performance Liquid Chromatography

A direct, isocratic reversed phase HPLC method was developed, based upon use of a symmetry C18 silica column (250 × 4.6 mm). The method uses a mixture of 0.2M ammonium acetate solution and acetonitrile (1:1 v/v) as the mobile phase, eluted at flow rate of about 1 mL/min. This method yielded the best resolution between Sildenafil and related substances, all of which were detected at a wavelength of 240 nm. The method validation showed Sildenafil and related substances were well resolved, indicative of good method specificity. Linearity for Sildenafil was found over the range of 0.05-0.2 mg/mL, and linearity for related substances was found over the range of 0.002-0.01 mg/mL. The method was found to be accurate, with the percentage bias being about 2% for Sildenafil and 4.5% for the related substances. In addition, the method was precise, showing a relative standard deviation of about 1% for both determination of Sildenafil and the related substances. The limit of quantitation was found to be in the range of 0.02-0.15 mg/100 mL. A typical chromatogram obtained using this method is shown in Figure 11 [6].

Figure 11.   Typical HPLC Chromatogram of Sildenafil citrate and its
              Related Substances.

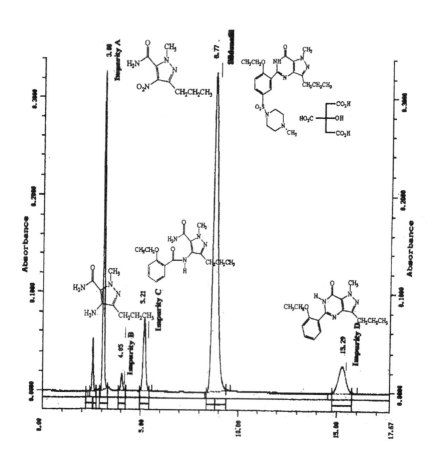

Another HPLC method for Sildenafil was reported, which is based on the use of a 4.6 mm x 10 cm column containing 5-μm L26 packing (butyl silane chemically bonded to totally porous silica particles). The mobile phase is 17: 7:1 water / pH 4.5 phosphate buffer / acetonitrile, eluted at a flow rate of 1.5 mL/min. Detection and quantitation is made on the basis of the UV absorbance at 230 nm.

4.5   Determination in Body Fluids and Tissues

4.5.1   HPLC Determination of Sildenafil and Metabolites in Plasma

A method suitable for the determination of Sildenafil citrate and its demethylated metabolite in blood plasma was developed using a kromasil C_4 (5μm, 100 × 4.6 mm I.D.) column. The mobile phase was acetonitrile / potassium phosphate buffer (500 mmol/L, pH 4.5) / water prepared at the ratio of 28:4:68, v/v/v, and eluted at a flow rate of 1.5 mL/min. 10 mmol/L of diethylamine hydrochloride was first dissolved in the buffer/water prior to addition of the acetonitrile. The temperature of the column was maintained at 40°C in a column block heater, and detection was made on the basis of the UV absorption at 230 nm.

Sample preparation was carried out using an ASTED (automated sequential trace enrichment of dailysates) system. This procedure is selective for both Sildenafil and its metabolite, and is linear over the range 1.0-250 ng/mL. 650 μL of plasma was mixed with 150 μL of MCA-reference compound, and 780 μL of this mixture delivered into the donor channel of the dialyzer. Recipient solvent (7000 μL) was moved through the TEC in a 6-minute time period. Following enrichment, the donor tubing and dialyzer channel were purged with 1500 μL of donor solvent (water) and 200 μL of 40% acetonitrile. Following the movement of 1000 μL of donor solvent (via UVSM value switching) through the TEC, the Rheodyne high pressure valve was switched to the inject position and the analyte eluted onto the HPLC column by the HPLC mobile phase [4].

4.5.1   ELISA Immunosorbant Assay

Enzyme-linked immunosorbant (ELISA) assay was used to detect and measure Sildenafil concentrations. Anti-Sildenafil antibody was developed in rabbits following the immunization by means of a drug conjugate with bovine serum albumin. The anti-Sildenafil antibody was then purified by ammonium sulfate, followed by protein A. The minimum concentration of Sildenafil that can be detected by the assay without interference is 20 ng/mL. The intra-and inter-assay percentage coefficients of variation noted for all standard points were less than 80% [7].

5.   Stability

5.1   Solid-State Stability

Sildenafil citrate in the solid-state was shown to be stable when stored for 2 months at conditions of 40°C / 75% RH, 50°C / 75% RH, and under exposure to UV light. The results are collected in Table 9 [3].

5.2   Solution-Phase Stability

In the solution phase, once Sildenafil was exposed to UV light for 2 months, a slight increase of impurities was encountered. This data is also found in Table 9 [3].

5.3   pH Stability Profile

The pH stability profile of Sildenafil citrate over the pH range of 1-13 was studied under severe storage condition (65°C for 2 weeks). The results showed high stability of Sildenafil citrate at the whole range, with no significant degradation being noticed. the results are plotted in Figure 12 [3].

5.4   Drug-Excipient Interactions

Sildenafil citrate was separately mixed with microcrystalline cellulose, calcium hydrogen phosphate, croscarmellose sodium, and magnesium stearate. The mixtures were then stored for one month at 40°C / 75% RH, 50°C / 75% RH, and at room temperature. Sildenafil citrate did not show any signs of instability indicating its compatibility with these common excipients [3].

Table 9

Stability of Sildenafil Citrate at Different Storage Conditions

Test required	Acceptable Limit	Initial Result	40°C / 75%RH	50°C / 75%RH	UV light	UV light; Solution Phase
			1 month	1 months	2 months	2 months
Appearance	White to off white crystalline powder	White to off white crystalline powder	White to off white crystalline powder	White to off white crystalline powder	White to off white crystalline powder	Off white crystalline powder
Assay	98.0%-102.0%	99.7%	100.7%	100.2%	99.2%	101.9%
Chromatographic Purity						
Any individual impurity:	< 0.5%	0.06%	0.06%	0.06%	0.13%	0.18%
Total impurities:	< 1.0%	0.07%	0.11%	0.11%	0.32%	0.52%

Figure 12.   pH stability profile for Sildenafil citrate over the pH range of 1-13.

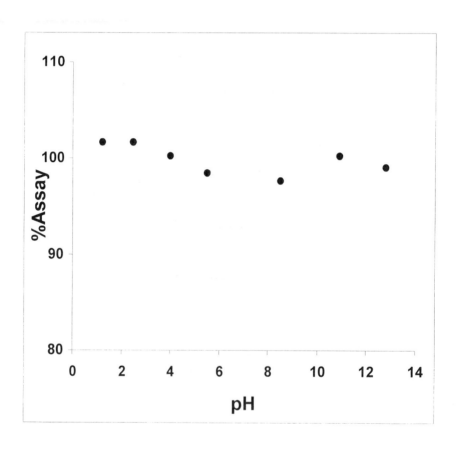

6.   Drug Metabolism and Pharmacokinetics

6.1   Absorption and Bioavailability

Sildenafil is rapidly absorbed after oral administration, having an absolute bioavailability of about 40%. Maximum observed plasma concentrations are reached within 30 to 120 minutes (median 60 minutes) after oral dosing in the fasted state. When Sildenafil is taken with a high fat meal, the rate of absorption is reduced, with a mean delay in T_{max} of 60 minutes and a mean reduction in C_{max} of 29% [1].

The drug has a duration of action lasting up to 4 hours, with less response than that seen at 2 hours [9, 10].

6.2   Distribution

The mean steady state volume of distribution (V_{ss}) for Sildenafil citrate is 105 L, indicating distribution into various tissues. Measurements of Sildenafil in the semen of healthy volunteers 90 minutes after dosing showed less than 0.001% of the administered dose located in their semen [9, 10].

Sildenafil and its major circulating N-desmethyl metabolite are approximately 96% bound to plasma proteins. The degree of protein binding is independent of the total drug concentration [9,10].

6.3   Metabolism and Elimination

The five principal pathways of metabolism found in mouse, rat, rabbit, dog, and man were piperazine N-demethylation, pyrazole N-demethylation, loss of a two-carbon fragment from the piperazine ring (N, N-deethylation), and oxidation of the piperazine ring and aliphatic hydroxylation. Additional metabolites arose through combinations of these pathways [11].

Sildenafil is cleared predominantly by the CYP3A4 (major route) and CYP2C9 (minor route) hepatic microsomal isoenzymes. The major circulating metabolite results from N-desmethylation of Sildenafil, which is further metabolized. This metabolite has a phosphodiesterase PDE selectivity profile similar to that of Sildenafil, and an *in vitro* potency for

PDE5 that is approximately 50% of the parent drug. Plasma concentrations of this metabolite are approximately 40% of those seen for Sildenafil, so that the metabolite accounts for about 20% of the pharmacological effects of Sildenafil.

The concomitant use of potent cytochrome P450 3A4 inhibitors (e.g., erythromycin and ketoconazole), as well as the nonspecific CYP inhibitor, cimetidine, is associated with increased plasma levels of Sildenafil. Both Sildenafil and the metabolite have terminal half-lives of about 4 hours.

After either oral or intravenous administration, Sildenafil is excreted predominantly in the feces as its metabolites (approximately 80% of administered oral dose), and to a lesser extent in the urine (approximately 13% of the administered oral dose). Similar values for pharmacokinetic parameters were seen in normal volunteers and in the patient population, using a population pharmacokinetic approach [1,9].

7.   Pharmacology

7.1   Mechanism of Action

The physiologic mechanism of erection of the male organ involves release of nitric oxide (NO) in the corpus cavernosum during sexual stimulation. Nitric oxide then activates the enzyme guanylate cyclase, which results in increased levels of cyclic guanosine monophosphate (cGMP), producing smooth muscle relaxation in the corpus cavernosum and allowing inflow of blood. Sildenafil has no direct relaxant effect on isolated human corpus cavernosum, but enhances the effect of nitric oxide by inhibiting phophodiesterase type 5 (PDE5), which is responsible for degradation of cGMP in the corpus cavernosum. When sexual stimulation causes local release of NO, inhibition of PDE5 by Sildenafil causes increased levels of cGMP in the corpus cavernosum, resulting in smooth muscle relaxation and inflow of blood to the corpus cavernosum. Sildenafil at recommended doses has no effect in the absence of sexual stimulation [1].

The advent of a new principle, amplification of the NO-signaling cascade by means of target organ selective phosphodiesterase inhibition, has renewed interest in phosphodiesterases and cGMP [12].

The NO-cGMP pathway also plays an important role in mediating blood pressure. It is therefore possible that the therapeutic doses of Sildenafil used to treat erectile dysfunction may have clinically significant effects on human hemodynamics [13].

7.2   Toxicity

7.2.1   Animals

Single oral dose studies in rats and mice located the minimal lethal dose to be between 500-1000 mg/kg in mice and between 300-500 mg/kg in rats. In rats, the severity of clinical signs in females and the mortality that occurred in females only suggested a sex-linked difference in the sensitivity to acute effects of Sildenafil free base.

Sildenafil was not carcinogenic when administered to rats for 24 months at a dose resulting in total systemic drug exposure (AUC) for unbound Sildenafil and its major metabolite of 29-and 42-times for male and female rats respectively, the exposure observed in human males given the Maximum Recommended Human Dose (MRHD) of 100 mg. Sildenafil was not carcinogenic when administered to mice for 18-21 months at dosages up to the Maximum Tolerated Dose (MTD) of 10 mg/Kg/day, approximately 0.6 times the MRHD on a mg/m^2 basis [1].

7.2.2 Humans

Adverse reactions were headache (16%), flushing (10%), dyspepsia (7%), nasal congestion (4%), urinary tract infection (3%), abnormal vision (3%), diarrhea (3%), dizziness (2%), and rash (2%). In fixed dose studies, dyspepsia (17%) and abnormal vision (11%) were more common at 100 mg than at lower doses. In studies with healthy volunteers taking single doses up to 800 mg, adverse events were similar to those seen at lower doses, but incidence rates were increased.

In cases of overdose, standard supportive measures should be adopted as required. Renal dialysis is not expected to accelerate clearance as Sildenafil is highly bound to plasma proteins and it is not eliminated in the urine [1].

References

1.   *Viagra®: (Sildenafil Citrate) the FDA Approved Impotence Pill*,
      Pfizer Labs., Division of Pfizer Inc., New York, 1999.

2.   N.K. Terrett, A.S. Bell, D. Brown, and P. Ellis, *Bioorg. Med.
      Chem. Lett.*, **6**, 1819 (1996).

3.   A.A. Badwan, N. Nabulsi, M. Al-Omari, and N. Daraghmeh,
      unpublished results, The Jordanian Pharmaceutical Manufacturing
      Company, P.O. Box 94, Naor 11710, Jordan.

4.   J.D. Cooper, D.C. Muirhead, J.E. Taylor, and R.P. Baker, *J.
      Chrom. B, Biomed. Sci. Appl.*, **701**, 87 (1997).

5.   I.M. Admour, "Solid State Modification and Transformation in
      Sildenafil and Terfenadine Salts", MSc. Thesis, Jordan University
      of Science and Technology, Irbid, Jordan, January 1999.

6.   N. Daraghmeh, M. Al-Omari, A.A. Badwan, and A.M.Y. Jaber,
      WHERE, in press.

7.   T. El-Thaher, "Sildenafil Analysis Kit File", Aragen Company,
      P.O. Box 94, Naor 11710, Jordan.

8.   *USP Pharmacopeial Forum*, **24**, 7185 (1998).

9.   M.M. Goldenberg, *Clin. Therap.*, **20**, 1033 (1998)

10.   *USP DI: Drug Information for Health Care Professional*, Vols. I
      & II update, The United States Pharmacopoeial Convention,
      Rockville, Maryland, 1998, p. 975.

11.   D.K. Walker, M.J. Achland, G.C. Jomes, G.J. Muirhead, D.J.
      Ramce, P. Wastall, and P.A. Wright, *Xenobioetica*, **29**, 297 (1999).

12.   H. Grossman, G. Petrischor, and G. Bartsch, *Exp. Gerontol.*, **34**,
      305 (1999).

13.   G. Jackson, N. Benjamin, N. Jackson, and M.J. Allen, *Am. J.
      Cardiol.*, **93**,13C (1999).

DORZOLAMIDE HYDROCHLORIDE

Marie-Paule Quint[1], Jeffrey Grove[1], and Scott M. Thomas[2]

(1) Laboratoires Merck Sharp & Dohme-Chibret
Centre de Recherche
Riom, France

(2) Merck Research Laboratories
Rahway, NJ
USA

ANALYTICAL PROFILES OF
DRUG SUBSTANCES AND EXCIPIENTS
VOLUME 27

377

The analytical profile on Dorzolamide Hydrochloride originally appeared
in Volume 26 of the *Analytical Profiles of Drug Substances and
Excipients*. Due to events beyond the authors' control, the following
typographical errors were not corrected before printing:

1. The title and page headers (pages 283-317) should read Dorzolamide
    Hydrochloride (and not Dorsolamide Hydrochloride).

2. Page 287, paragraph 2, line 12 read ethoxzolamide (not
    ethoxzolaniide).

3. Page 296, Figure 4, the units for the ordinate should read solubility
    (mg/mL, as base).

4. Page 299, section 3.7.2, read Table 3 (not Table 1).

5. Page 304, read Table 4 (not Table 5).

6. Page 312, the figures showing the structures of the two degradation
    products should be at the end of section 5.2 and not 5.1. Dorzolamide
    is stable in the solid state (5.1), degradation occurs in the liquid state
    (5.2).

7. Page 315 read [17] at paragraph end (not [16]).

MESALAMINE

Alekha K. Dash[1] and Harry G. Brittain[2]

(1) Department of Pharmaceutical & Administrative Sciences
School of Pharmacy and Allied Health Professions
Creighton University
Omaha, NE 68178
USA

(2) Center for Pharmaceutical Physics
10 Charles Road
Milford, NJ 08848
USA

ANALYTICAL PROFILES OF
DRUG SUBSTANCES AND EXCIPIENTS
VOLUME 27

379

The analytical profile on Mesalamine originally appeared in Volume 25 of the *Analytical Profiles of Drug Substances and Excipients*. Table 6 was omitted from the published monograph, and is reproduced here:

Table 6

Summary of the Reported HPLC Methods Used for the Analysis of 5-Aminosalicylic Acid

Method Category	Mobile Phase Composition	Special agent used	Column	Detector	Reference
Aqueous samples, reversed phase method	pH 4.8 ammonium phosphate (0.05 M); Flow = 1 mL/min	none	C18 (5mm): 10 cm x 4.6 mm (id)	Electrochemical/ Photodiode array	28
Aqueous or formulation samples, reversed phase method	Methanol:phosphate buffer (pH 7.4)(20:80 v/v); Flow = 1.0 mL/min	none	C8 (5mm): 25 cm x 4.6 mm (id)	UV (290 nm)	31
Aqueous samples, ion pairing method	Methanol:phosphate buffer (70:30 v/v); Flow = 1.8 mL/min	tetrabutyl ammonium	C18 (5mm): 15 cm x 4.6 mm (id)	UV (254 nm)	26
Formulation samples, reversed phase method	Acetonitrile:Water (22:78 v/v) with 0.5% Acetic acid; Flow = 1 mL/min	none	C18: 25 cm x 4 mm (id)	UV (300 nm)	27

Sample/Method	Mobile Phase	Reagent	Column	Detection	Ref
Formulation samples, ion pairing method	Acetonitrile:0.1 M potassium phosphate buffer:water (18:10:72 v/v); Flow = 1.5 mL/min	tetrabutyl ammonium	C18 (5mm): 12 cm x 4.6 mm(id)	Photodiode array (240, 310, 410 nm)	29
Biological fluid samples, derivitization method	0.05 M sodium sulfate buffer:acetonitrile (1:1 v/v); Flow = 1.3 mL/min	benzyl chloroform ate	Anion exchange: (10mm)25 cm x 4 mm(id)	Fluorescence: (EX = 300 nm EM = 470 nm)	25
Biological fluid samples, ion pairing method	Acetonitrile:0.05 M Potassium dihydrogen Phosphate solution (15:85 v/v)	0.1% tetrabutyl ammonium hydroxide	LiChrosorb 10 RP C18 (Merck)	Fluorescence (EX = 360 nm EM = 425 nm)	21
Biological fluid samples, reversed phase method	Methanol: Buffer (15:85 v/v) Buffer consisted of 0.12 mmol/L sodium octylhydrogen sulfate, 0.05 mmol/l EDTA, 0.2 mol/L citric acid, and 0.01 mol/L Na_2HPO_4 solution. pH adjusted to 3 with H_3PO_4; Flow = 0.8 mL/min	none	Spherisorb C18-2: (25 cm x 4.6 mm)	Electrochemical detector	24

CUMULATIVE INDEX

Bold numerals refer to volume numbers.

ISBN 0-12-260827-5

90065